Adaptive Control of
Underactuated
Mechanical
Systems

Adaptive Control of
Underactuated
Mechanical
Systems

An-Chyau Huang
Yung-Feng Chen
Chen-Yu Kai

National Taiwan University of Science and Technology, Taiwan

NEW JERSEY • LONDON • SINGAPORE • BEIJING • SHANGHAI • HONG KONG • TAIPEI • CHENNAI

Published by

World Scientific Publishing Co. Pte. Ltd.
5 Toh Tuck Link, Singapore 596224
USA office: 27 Warren Street, Suite 401-402, Hackensack, NJ 07601
UK office: 57 Shelton Street, Covent Garden, London WC2H 9HE

British Library Cataloguing-in-Publication Data
A catalogue record for this book is available from the British Library.

ADAPTIVE CONTROL OF UNDERACTUATED MECHANICAL SYSTEMS

ISBN 978-981-4663-54-0

Printed in Singapore

To Our Families

Preface

The aim of this book is to address recent developments of the adaptive control for a class of single input underactuated mechanical systems. A mechanical system is underactuated if it is driven by fewer actuators than its degrees of freedom. Since some degrees of freedoms are not directly driven by the actuators, the control of underactuated systems is generally challenging. However, many real life systems are underactuated, and hence the study of their controller design is very important. It is known that the underactuated dynamics for each underactuated system is unique and a specific control scheme is needed to stabilize the closed-loop system. Unlike fully actuated systems where many control strategies are available for an entire class of systems, there are few results are applicable to even a small class of underactuated mechanical systems. In this book, we would like to introduce a possible way to design a strategy to control a class of single input underactuated systems which may contain uncertainties and/or subject to external disturbances. A diffeomorphism is found to decouple the underactuated dynamics so that the system can be represented into a special cascade form. Presence of the mismatched general uncertainties in this cascade form prevents the application of most conventional designs. The main tool used in this book is the function approximation technique which represents the general uncertainties as finite combinations of basis functions weighted with unknown constant coefficients. The adaptive multiple-surface sliding control is used to deal with the mismatched uncertainties. The same controller has been used to control several benchmark underactuated systems, and simulation results show satisfactory performance.

The book has been written as a text suitable for graduate students in the advanced course for the control of underactuated systems. It is also intended to provide valuable tools for researches and practicing engineers who currently work on the control of underactuated mechanical systems.

We would like to thank all of our colleagues and students with whom we have discussed the basic problems in the control of underactuated systems over the last few years. Without them, many issues would never have been clarified. This work was partially supported by the Ministry of Science and Technology of the Republic of China government and by the National Taiwan University of Science and Technology. The authors are grateful for their supports.

An-Chyau Huang,
Yung-Feng Chen,
Chen-Yu Kai

Mechanical Engineering Department
National Taiwan University of Science and Technology

Contents

Chapter 1

Introduction

Underactuated systems are mechanical systems with fewer actuators than the system degrees of freedom, i.e., some degrees of freedoms are not directly driven by actuators (Spong, 1997; Liu and Yu, 2013). Many real life systems are underactuated. Examples include the road vehicles, mobile robots, underwater vehicles, surface vessels, helicopters, and some robot manipulators. Since there are fewer actuators available to drive the system, controllability of underactuated systems is very important. To have an effective control, this book considers controllable underactuated systems only.

Nonholonomic Constraints and Underactuated Systems

A lot of underactuated mechanical systems contain nonholonomic acceleration constraints. A nonholonomic constraint is a constraint equation of a mechanical system that cannot be written as a time derivative of some function of the generalized coordinates, and hence not integrable. There are two main nonholonomic constraints in classical mechanics for mechanical systems: the velocity nonholonomic constraints and acceleration nonholonomic constraints. The former can be written as a function of the generalized coordinate $\mathbf{q}$ and the generalized velocity $\dot{\mathbf{q}}$ in the form of $\boldsymbol{\varphi}(\mathbf{q},\dot{\mathbf{q}}) = \mathbf{0}$ that are not integrable. In addition to $\mathbf{q}$ and $\dot{\mathbf{q}}$, the latter contains the generalized acceleration like $\boldsymbol{\varphi}(\mathbf{q},\dot{\mathbf{q}},\ddot{\mathbf{q}}) = \mathbf{0}$ that are not integrable either. In this book, we assume that all systems considered have nonholonomic acceleration constrains that will be detailed in chapter 3.

Linear/Nonlinear Underactuated Systems

Not all underactuated systems are nonlinear, although most of the underactuated systems considered in the literature exhibit nonlinear dynamics. For example, in an active control of a quarter-car vehicle suspension system shown in Fig. 1.1, the sprung mass m_s represents the car body and the unsprung mass m_u is an assembly of the axle and wheel. There is only one actuator force u in between the sprung and unsprung mass to control both the car body displacement x_1 and the tire displacement x_3 when subject to the road profile excitation z. Since all the dampers and springs are linear, this is a linear underactuated system.

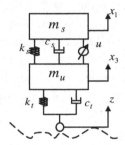

Figure 1.1.   A quarter-car vehicle model

Simple Control by Ignoring the Underactuated Dynamics

Although the control of underactuated systems are generally challenging, some special systems can be controlled by ignoring the underactuated dynamics. For example, a lot of control strategies designed for the quarter-car suspension system in Fig. 1.1 focus on the regulation of the sprung mass while regarding the unsprung mass behavior as a stable internal dynamics due to its dissipative nature. This design largely reduces the complexity of the control problem. Therefore, an intuitive way for controlling an underactuated system can possibly be the control of the actuated part and ignore the unactuated dynamics. However, most of the underactuated systems do not have stable unactuated subsystem, and this simple design may not be widely applicable. For example, we may also see some literature on the control of cart pole systems in such a way that the pendulum was regulated to the vertical position by the controller while the cart was driven to infinity, i.e., the cart dynamics is unbounded.

Feedback Linearization and Underactuated Systems

Feedback linearization is an effective approach for controlling nonlinear systems when the system model contains no uncertainties. However, it can be easily shown that nonlinear underactuated systems are not exact feedback linearizable. Although the well-known partial feedback linearization can be applied to transform the actuated dynamics into a double integrator, the unactuated subsystem generally still exhibits a complex dynamics. Some effective control strategies are needed to obtain desired performance. One way to eliminate the internal dynamics, if possible, is to redefine the output function (Slotine and Li, 1991) when applying the input/output feedback linearization. For example, let us consider the quarter-car vehicle suspension in Fig. 1.1 again with the linear dynamic equation

$$\dot{x}_1 = x_2$$
$$\dot{x}_2 = \frac{1}{m_s}[k_s(x_3 - x_1) + c_s(x_4 - x_2) + u]$$
$$\dot{x}_3 = x_4 \tag{1}$$
$$\dot{x}_4 = \frac{1}{m_u}[k_s(x_1 - x_3) + c_s(x_2 - x_4) - u + k_t(z - x_3) + c_t(\dot{z} - x_4)]$$

Instead of selecting the output $y = x_1$, we redefine it to be $y = x_1 + rx_3$ where $r > 0$ is a weighting factor. By taking the time derivative of the output twice, we have

$$\ddot{y} = \ddot{x}_1 + r\ddot{x}_3$$
$$= \frac{1}{m_s}[k_s(x_3 - x_1) + c_s(x_4 - x_2) + u] \tag{2}$$
$$+ \frac{r}{m_u}[k_s(x_1 - x_3) + c_s(x_2 - x_4) - u + k_t(z - x_3) + c_t(\dot{z} - x_4)]$$

We see that the same control coming from the two subsystems appears in (2). This implies that it could be possible to select this u to regulate the two subsystems at the same time. After some straightforward rearrangement, we obtain the dynamic equation affine in the control signal as

$$\ddot{y} = f + gu \tag{3}$$

where

$$f = \frac{1}{m_s}[k_s(x_3 - x_1) + c_s(x_4 - x_2)]$$

$$+ \frac{r}{m_u}[k_s(x_1 - x_3) + c_s(x_2 - x_4) + k_t(z - x_3) + c_t(\dot{z} - x_4)]$$

$$g = \frac{1}{m_s} - \frac{r}{m_u}$$

Suppose both z and $\dot{z}$ are known, all system parameters are available and r is selected so that $rm_s \neq m_u$, then we may design the controller to be

$$u = \frac{1}{g}(-f - k_v\dot{y} - k_p y) \tag{4}$$

where k_v and k_p are positive gains so that the closed-loop dynamics

$$\ddot{y} + k_v\dot{y} + k_p y = 0 \tag{5}$$

ensures $y \to 0$ as $t \to \infty$. However, convergence of output signal y does not imply convergence of the car body displacement x_1 or tire displacement x_3. This is because $y = x_1 + rx_3 \to 0$ and the system states converge to the line $x_1 + rx_3 = 0$ in the state space instead of the origin. Generally, adjusting of the weighting factor r may not always give significant improvement of the control performance in this approach. Hence, the output redefinition fails in this application.

Benchmark Underactuated Systems

Since most real life underactuated mechanical systems contain complex dynamics, it is not easy to clarify the impact of the underactuated structure on the entire system behavior. Many benchmark underactuated systems have then been developed to study the nature of underactuated dynamics and their possible controller designs. These systems include, but are not limited to, the cart pole system, overhead crane, TORA, rotary inverted pendulum, acrobot, and pendubot. Each one of these systems is with unique underactuated dynamics, and a specific design is needed. Unlike fully actuated systems where many control strategies are available for an entire class of systems, there are few results are applicable to even a small class of underactuated mechanical systems (Spong, 1997).

Control of a Class of Underactuated Systems

In this book, we would like to introduce a possible way to design a strategy to control a class of single input underactuated systems which may contain uncertainties and/or subject to external disturbances. The basic idea comes from the fact that each subsystem in a mechanical system can be represented as a second order differential equation which is also valid to underactuated mechanical systems. For controllable single input underactuated systems, the control effort is shared by each and every subsystem in the form

$$
\begin{aligned}
\left.\begin{aligned}
\dot{x}_1 &= x_2 \\
\dot{x}_2 &= f_2(\mathbf{x}) + b_2(\mathbf{x})u
\end{aligned}\right\} &\text{ subsystem 1} \\[6pt]
\left.\begin{aligned}
\dot{x}_3 &= x_4 \\
\dot{x}_4 &= f_4(\mathbf{x}) + b_4(\mathbf{x})u
\end{aligned}\right\} &\text{ subsystem 2} \\[6pt]
\vdots \qquad\qquad& \\[6pt]
\left.\begin{aligned}
\dot{x}_{n-1} &= x_n \\
\dot{x}_n &= f_n(\mathbf{x}) + b_n(\mathbf{x})u
\end{aligned}\right\} &\text{ subsystem } \tfrac{n}{2}
\end{aligned}
\tag{6}
$$

It is seen that the same u appears in all subsystems represented by second order dynamics. The system is extremely difficult to control because the control effort has to stabilize all subsystems at the same time. One way to circumvent the difficulty is to decouple all the subsystems from the control effort except the last one. This can be achieved by finding a diffeomorphism using similar derivation of coordinate transformations in traditional feedback linearization designs. After the transformation, the system may be represented in a new space as

$$
\begin{aligned}
\dot{z}_1 &= z_2 + d_1(\mathbf{z}) \\
\dot{z}_2 &= z_3 + d_2(\mathbf{z}) \\
&\vdots \\
\dot{z}_{n-3} &= z_{n-2} + d_{n-3}(\mathbf{z}) \\
\dot{z}_{n-2} &= z_{n-1} + d_{n-2}(\mathbf{z}) \\
\dot{z}_{n-1} &= z_n + d_{n-1}(\mathbf{z}) \\
\dot{z}_n &= d_n(\mathbf{z}) + b_n(\mathbf{z})u
\end{aligned}
\tag{7}
$$

Therefore, there is only one u appears in the last equation and the whole system is in a special cascade form where the i-th equation is connected to the next equation with signal z_{i+1}. Since $d_i(\mathbf{z})$, $i = 1,...,n-1$ are functions of the state vector, the system after transformation does not in general satisfy the pure feedback requirement and hence the well-known backstepping procedure is not applicable. In addition, during the coordinate transformation, these $d_i(\mathbf{z})$, $i = 1,...,n-1$ might be too complex to calculate for some specific systems. Although it could be possible to complete these computations by using symbolic manipulation tools in software packages like Matlab and Mathematica, the forms of their results may not always be suitable to include into the equations of motion directly. In this book, we would like to assume these complex terms to be uncertainties so that there is no need to compute them explicitly. Since these uncertainties enter the system in a mismatched fashion, few methods are feasible to cover their effects. Because we allow these uncertainties in the transformed domain, this is also equivalent to say that we permit some corresponding uncertainties in the original system.

Stabilization of an underactuated system is generally difficult, but we would like to make the problem more challenging by requiring the controller to be able to stabilize a class of single input uncertain underactuated system as shown in (6). We are going to find a diffeomorphism first to transform (6) into (7) to decouple the underactuated dynamics. A controller is then designed to stabilize the system in the special cascade form in (7) regardless of various mismatched uncertainties.

Control of Uncertain Underactuated Systems

Two main approaches are available for dealing with uncertainties in control systems. The robust strategies need to know the worst case of the system so that a fixed controller is able to be constructed to cover the uncertainties. In most cases, the worst case of the system is evaluated by proper modeling of the uncertainties either in the time domain or frequency domain. The variation bounds estimated from the uncertainty model are then used to design the robust terms in the controller. In some practical cases, however, these variation bounds are not available, and hence most robust strategies are infeasible. The other approach for dealing with system uncertainties is the adaptive method. Although intuitively we think that an adaptive controller should be able to give good performance to a system with time-varying uncertainties, conventional adaptive designs can actually be useful to systems with constant uncertainties. Therefore, to be feasible for the adaptive designs all time-varying parts in the system dynamics should be collected into a known regressor matrix, while the unknown constant parameters are put into a parameter vector. This process is called the linear parameterization of the uncertainties which is almost a must for adaptive designs. After the parameterization, proper update laws can then be derived to provide sufficient information to the certainty equivalence based adaptive controller such that the closed loop system can give good performance. However, there are some practical cases whose uncertainties are not able to be linearly parameterized (e.g., various friction effects), and some others are linearly parameterizable but the regressor matrices are too complex to derive (e.g., robot manipulators).

In our approach for the control of underactuated systems, the uncertainties in (7) are assumed to be time-varying without knowing their variation bounds. Due to their time-varying nature, most traditional adaptive designs are not feasible. Since their variation bounds are not known, most conventional robust schemes are not applicable. In this book, we are going to call this kind of uncertainties the *general uncertainties* which will be detailed in chapter 2. For a system with general uncertainties, few control schemes are available to stabilize the closed loop dynamics.

Function Approximation Based Controller Designs

The main strategy we are going to use in this book to deal with the uncertainties is based on the function approximation technique (FAT) proposed by Huang and Kuo (2001). The basic idea of the technique is to represent the uncertain terms into a finite combination of known basis functions. This effectively transforms a general uncertainty into a known basis vector multiplied by a vector of unknown coefficients. Since these coefficients are constants, update laws can be derived by using the Lyapunov-like methods. Due to the fact that the mathematical background for the function approximation has well been established and the controller design portion follows the traditional adaptive strategies, the function approximation based adaptive method provides an effective tool in dealing with controller design problems involving the general uncertainties. This method has been applied to various systems such as friction compensation (Alamir, 2002), robot manipulators (Chien and Huang, 2004; Huang et al., 2006; Huang and Chien, 2010; Azlan and Yamaura, 2012, 2013; Kai and Huang, 2013a; Talaei et al., 2013; Li et al., 2013; Kai and Huang, 2014a; Al-Shuka et al., 2014), flexible joint robots (Huang and Chen, 2004a; Chien and Huang, 2007, 2009, 2010a, 2011), jet engine control (Tyan and Lee, 2005), active vehicle suspensions (Chen and Huang, 2005a, 2005b, 2006), flexible link robots (Huang and Liao, 2006), vibration control (Chang and Shaw, 2007), pneumatic servo (Tsai and Huang, 2008a, 2008b), DC motors (Liang et al., 2008; Cong et al., 2009), visual servo (Chien and Huang, 2010b), belt-driven servo (Lee and Huang, 2011), vibration absorbers (Kai and Huang, 2013), and active linearization (Kai and Huang, 2013).

The history of the function approximation technique dates back to hundreds of years ago when Joseph Fourier initiated the investigation of heat transfer problems by using orthogonal function expansions. Therefore, the method is even older than the control theory itself. Many background materials of the function approximation technique are reviewed in chapter 2 for convenience.

Multiple-Surface Sliding Control and Mismatched Uncertainties

Sliding control is an effective robust scheme for the control of a class of nonlinear systems with uncertainties defined in compact sets. At any moment, the direction of the control action is determined by the switching condition to force the system to evolve on the switching surface so that the closed-loop dynamics behaves like a stable lower order linear system and the system states can converge to the target set according to the dynamics of the surface. For the method to be feasible, the uncertainties should be in the range space of the control input so that during the sliding mode the system behavior maintains an invariance property, which is independent of the uncertainties. This requirement is well known as the *matching condition* (Barmish and Leitmann, 1982; Chen and Leitmann, 1987). Many practical control systems, however, contains uncertainties that do not satisfy the matching condition. If the traditional sliding controller is applied to these systems, the mismatched uncertainties may enter into the system dynamics and destabilize the sliding mode. A trivial method to solve the present problem is to require that the bounds on the derivatives of the uncertainties be available. Since we are only given the bounds of the uncertainties themselves in most real life problems, it is impractical to have any further requirement on the bounds of the derivatives of the uncertainties.

The most well-known strategy for dealing with mismatch uncertainties is the backstepping procedure. However, our approach for underactuated system will end up with a special cascade system (7) not in the pure feedback structure and the backstepping may not be applicable. Therefore, we resort to the multiple-surface sliding control proposed by Green and Hedrick (1990) which is also very effective in coping with the mismatched uncertainties. Many practical applications of the multiple-surface sliding control have been implemented successfully (Green and Hedrick, 1990; McMahon et al., 1990; Stotsky et al., 1997; Gerdes and Hedrick, 1997). This method allows a sliding controller to be designed as if the system has a reduced relative degree by defining of the states as a synthetic input to the reduced order plant. The synthetic states are then controlled by a second controller to make sure that the synthetic control will track the profile defined by the sliding controller. One difficulty for the implementation of this approach is to calculate the differentiation of the desired trajectories of the synthetic input. In Green and Hedrick (1990), it was approximated by first-order finite differences. Won and Hedrick (1996) proposed a multiple-surface sliding controller for a class of SISO nonlinear systems whose uncertainties do not satisfy the matching condition but satisfy a coordinate dependence condition on the uncertainty bounds. Stotsky et al. (1997) used sliding mode filters to estimate the derivative of the desired trajectory and applied the method to control system with non-Lipschitz nonlinearities. Gerdes and Hedrick (1999) proposed a dynamic surface control scheme to integrate the filter into the system structure to remove the need for numerical differentiation and relax certain smoothness assumptions.

FAT-based Adaptive Multiple-Surface Sliding Control

An FAT-based sliding controller was proposed in Huang and Kuo (2001) for a class of non-autonomous systems with matched general uncertainties. The key idea was to use the function approximation technique to transform the uncertain terms into a finite combination of orthonormal basis functions. Since the coefficients of the approximation series are time-invariant, the update law can be obtained from the conventional Lyapunov design. Huang and Chen (2004a) extended the method to integrate the FAT with the multiple-surface sliding control to deal with mismatched general uncertainties. This method was justified in Huang and Chen (2004b) with experimental studies for the control of a flexible joint robot subject to mismatched uncertainties. Chen and Huang (2005) applied the method to the control of hydraulic active suspension systems with satisfactory performance. Tsai and Huang (2008) designed an FAT-based multiple-surface sliding control for pneumatic servo systems with experimental verification. This method had also been applied to perform oscillation suppression in belt-driven servo systems (Lee and Huang, 2011). It is also very effective in velocity regulation of the brushless DC motor with unabridged model subject to load variations (Kai and Huang, 2013).

In controlling the class of single input underactuated systems in (6), we decouple the underactuated dynamics by using a coordinate transformation to give a system in the special cascade form in (7). The mismatched general uncertainties invalidate most of the control strategies. In this book, we would like to use the FAT-based adaptive multiple-surface sliding control to stabilize the system in (7) so that the original system (6) can also give desired performance.

Organization of the Book

In chapter 2, the mathematical preliminaries and control theories useful in this book are reviewed. Readers familiar to these fundamentals are suggested to go directly to the next chapter. Various concepts from the linear algebra and real analysis are briefly presented in this chapter. Some emphasis will be placed on the spaces where the function approximation techniques are valid. Various orthonormal functions are also listed with their effective ranges for the convenience in the selection of basis functions for the FAT-based designs. Then the Lyapunov stability theory and the Lyapunov-like methods are reviewed in detail followed by the introduction of the Frobenius theorem which is very useful in verifying existence of the coordinate transformation. Next, we review the sliding control and the model reference adaptive control theories to elucidate the usefulness and limitation for these two classes of strategies. Some schemes to robustify the adaptive loop is also introduced in this chapter. Then, the concept of the general uncertainties is detailed, and the FAT-based adaptive controller is designed to deal with these uncertainties. Finally, the backstepping procedure and the multiple-surface sliding control are presented for clarifying their usefulness in covering the mismatched uncertainties.

In chapter 3, we study the dynamics of underactuated systems and find a diffeomorphism to transform the system into a special cascade form. Firstly, we derive dynamic equations for general mechanical systems. Then, the underactuated systems are rigorously defined, and some of their properties are introduced. The collocated partial feedback linearization is derived to transform the actuated part of the system dynamics into a double integrator. The general coordinate transformation is derived for decoupling the underactuated dynamics. A simple transformation for a 2-DOF underactuated mechanical system is presented followed by its extension to a class of higher-order systems.

In chapter 4, the FAT-based adaptive multiple-surface sliding controller is derived to stabilize the decoupled system represented in the special cascade form. The control problem is firstly formulated with the investigation of the system model obtained in chapter 3 more closely. The controller is then derived with rigorous proof of closed-loop stability as well as the transient performance. It is shown that if the approximation error is ignorable, the asymptotic stability of the output error and boundedness of the internal signals can be obtained. When considering the approximation error, the output error is guaranteed to be ultimately uniformly bounded and the transient response is bounded by an exponential function.

From chapters 5 to 10, we will introduce the control of the benchmark systems respectively with cart-pole system, overhead crane, TORA, rotary inverted pendulum, vibration absorber, and pendubot. In these chapters, the specific system dynamics is derived first to elucidate its underactuated dynamics. Then the controller derived in chapter 4 is used to control the system. Simulation results are presented to justify its effectiveness. From these chapters, we know that the controller is able to stabilize a class of underactuated systems which is rarely seen in the literature.

Chapter 2

Preliminaries

Various background materials useful in this book are reviewed in this chapter. Most of the results are presented without proof since they can be found in most of the related textbooks. In the first section, we review some mathematical background such as the notions of various spaces, the best approximation problems in the Hilbert space, the concept of orthogonal functions and their applications, the definitions of normed function spaces, and representations of functions, vectors and matrices in function approximation applications. The second section reviews the Lyapunov stability theory which includes the definition of stabilities of equilibrium points in the sense of Lyapunov for autonomous and non-autonomous systems, Lyapunov stability theorems, invariant set theorem, and Barbalat's lemma. In section 2.3, some notions from differential geometry are introduced to facilitate the understanding of the Frobenius theorem which is very useful in concluding solvability for a set of partial differential equations. This theorem plays an important role in chapter 3 to determine existence of solution for decoupling of underactuated dynamics. Some preliminaries in control theories are also given in this chapter as the background for the theoretical development introduced in the later chapters. The concept of sliding control is provided in section 2.4 and some designs of adaptive strategies are presented in section 2.5. Robustness of the adaptive loop is reviewed in section 2.6 with traditional schemes such as dead-zone and σ-modification. Section 2.7 introduces the concept of general uncertainties. Due to the limitations in the traditional robust control and conventional adaptive design, it is challenging to control a system with general uncertainties. In section 2.8, we introduce the function approximation based adaptive controller for dealing with systems containing general uncertainties. When the uncertainties enter the system dynamics in a mismatched fashion, few strategies are feasible. The backstepping procedure is well-known to be effective in coping with mismatched uncertainties. We review the design method of the backstepping procedure in section 2.9. A backstepping-like design called multiple-surface sliding control is derived in section 2.10 which is the main method in this book to cover the general uncertainties entering the system dynamics in a mismatched fashion.

2.1 Review of Some Mathematical Background

Vector Spaces

Function approximation techniques play an important role in the design of adaptive controllers in this book. Let us review their mathematical foundations starting from the definition of vector spaces. A nonempty set X is a (*real*) *vector space* if the following axioms are satisfied. For any $\mathbf{x}, \mathbf{y} \in X$, the addition is closed as $\mathbf{x} + \mathbf{y} \in X$ and it is also commute like $\mathbf{x} + \mathbf{y} = \mathbf{y} + \mathbf{x}$. If $\mathbf{z}$ is a member of X, then the associative law of addition has to be satisfied as $(\mathbf{x} + \mathbf{y}) + \mathbf{z} = \mathbf{x} + (\mathbf{y} + \mathbf{z})$. There is a unique vector $\mathbf{0} \in X$ such hat $\mathbf{x} + \mathbf{0} = \mathbf{x}$ for all $\mathbf{x} \in X$. There exists a unique vector $-\mathbf{x} \in X$ such that $\mathbf{x} + (-\mathbf{x}) = \mathbf{0}$ for all $\mathbf{x} \in X$. Multiplication of any $\mathbf{x} \in X$ with $\alpha \in \Re$ is still in X, i.e. $\alpha \mathbf{x} \in X$. For $\beta \in \Re$, it further needs to satisfy $\alpha(\beta \mathbf{x}) = (\alpha \beta)\mathbf{x}$ and $(\alpha + \beta)\mathbf{x} = \alpha \mathbf{x} + \beta \mathbf{x}$. For all $\mathbf{x} \in X$, multiplication to the unitary element has to be invariant, i.e., $1\mathbf{x} = \mathbf{x}$. The distributive law $\alpha(\mathbf{x} + \mathbf{y}) = \alpha \mathbf{x} + \alpha \mathbf{y}$ holds for all $\mathbf{x}, \mathbf{y} \in X$ and $\alpha \in \Re$. For example, the set of all n-tuples of real numbers is a real vector space and is known as $\Re^n$. The set of all real-valued functions defined over an interval $[a, b] \in \Re$ is also a vector space. The set of all functions mapping the interval $[a, b] \in \Re$ into $\Re^n$ can also be proved to be a vector space. In some literature, the real vector space is also known as the *real linear space*.

If $\mathbf{x}_1, ..., \mathbf{x}_k \in \Re^n$ and $c_1, ..., c_k \in \Re$, a vector of the form $c_1 \mathbf{x}_1 + ... + c_k \mathbf{x}_k$ is called a *linear combination* of the vectors $\mathbf{x}_1, ..., \mathbf{x}_k$. The set of vectors $\mathbf{x}_1, ..., \mathbf{x}_k \in \Re^n$ is said to be *independent* if the relation $c_1\mathbf{x}_1 + ... + c_k\mathbf{x}_k = 0$ implies $c_i = 0, i = 1,...,k$; otherwise, the set of vectors is *dependent*. If $S \subset \mathfrak{R}^n$ and if R is the set of all linear combinations of elements of S, then S spans R, or we may say that R is the *span* of S. An independent subset of a vector space $X \subset \mathfrak{R}^n$ which spans X is called a *basis* of X. A vector space is *n-dimensional* if it contains an independent set of n vectors, but every set of $n+1$ vectors is a dependent set.

Metric Spaces

A set X of elements p, q, r,... is said to be a *metric space* if with any two points p and q there is associated a real number $d(p,q)$, the distance between p and q, such that $d(p,q) > 0$ if $p \neq q$; $d(p,p) = 0$; $d(p,q) = d(q,p)$; and for any $r \in X$ $d(p,q) \leq d(p,r) + d(r,q)$. The distance function on a metric space can be thought of as the length of a vector; therefore, many useful concepts can be defined. A set $E \subset X$ is said to be *open* if for every $x \in E$, there is a ball centered at x with radius r denoted as $B(x,r) = \{y \in X \,|\, d(y,x) < r\}$ such that $B(x,r) \subset E$. A set is *closed* if and only if its complement in X is open. A set E is *bounded* if $\exists r > 0$ such that $d(x,y) < r \,\forall x, y \in E$. Let $S \subset T$ and T is a subset of X. S is said to be *dense* in T if for each element t in T and each $\varepsilon > 0$, there exists an element s in S such that $d(s,t) < \varepsilon$. Thus every element of T can be approximated to arbitrary precision by elements of S.

Let X and Y be metric spaces with distance functions d_X and d_Y, respectively. A function $f : X \rightarrow Y$ is said to be *continuous* at a point $x \in X$ if $f(x + \delta x) \rightarrow f(x)$ whenever $\delta x \rightarrow 0$. Or, we may say, f is continuous at x if given $\varepsilon > 0$, there exist $\delta > 0$ such that $d_X(x,y) < \delta \Rightarrow d_Y(f(x),f(y)) < \varepsilon$. A function f is continuous on $E \subset X$ if it is continuous at every point of E, and it is *uniformly continuous* on E if given $\varepsilon > 0$, there exist $\delta(\varepsilon) > 0$ such that $d_X(x,y) < \delta \Rightarrow d_Y(f(x),f(y)) < \varepsilon$ for all $x, y \in E$.

The distance function generates the notion of convergence: A sequence $\{x_i\}$ in a metric space X is said to be *convergent* to an element x in X if for each $\varepsilon > 0$ there exists an integer n such that $d(x,x_i) \leq \varepsilon$ whenever $i > n$. A set $E \subset X$ is *compact* if each sequence of points in E

contains a subsequence which converges to a point in E. In particular, a compact subset of $\Re^n$ is necessarily closed and bounded. The sequence $\{x_i\}$ is a *Cauchy sequence* if for each $\varepsilon > 0$ there exists an integer n such that $p, q > n \Rightarrow d(x_p, x_q) \leq \varepsilon$. Clearly, every convergent sequence is a Cauchy sequence, but the converse is not true in general. A *complete* metric space X is a space where every Cauchy sequence converges to a point in X.

Normed Vector Spaces

Let X be a vector space, a real-valued function $\|\cdot\|$ defined on X is said to be a *norm* on X if it satisfies the following properties: For all $\mathbf{x} \in X$ and $\mathbf{x} \neq \mathbf{0}$, $\|\mathbf{x}\| > 0$; $\|\mathbf{0}\| = 0$; for any scalar α we have $\|\alpha\mathbf{x}\| = |\alpha|\|\mathbf{x}\|$; and $\|\mathbf{x} + \mathbf{y}\| \leq \|\mathbf{x}\| + \|\mathbf{y}\|$ for all $\mathbf{x}, \mathbf{y} \in X$. A *normed vector space* represented as the pair $(X, \|\cdot\|)$ is a metric space with the metric defined by $d(\mathbf{x}, \mathbf{y}) = \|\mathbf{x} - \mathbf{y}\|$, $\forall \mathbf{x}, \mathbf{y} \in X$. The concept of sequence convergence can be defined using the norm as the distance function. Hence, we are now ready to define convergence of series. The series $\sum \mathbf{x}_i$ is said to converge to a point $\mathbf{x} \in X$ if the sequence of partial sums converges to $\mathbf{x}$, i.e., if $\forall \varepsilon > 0, \exists n > 0$ such that $\left\| \sum_{i=1}^{m} \mathbf{x}_i - \mathbf{x} \right\| < \varepsilon$ whenever $m > n$.

A complete normed vector space is called a *Banach space*. In a normed vector space, the length of any vector can be defined by its norm. To define the angle between any two vectors, in particular the concept of orthogonality between vectors, we need the notion of the inner product space.

Inner Product Spaces and Hilbert Spaces

An *inner product* $\langle \mathbf{x}, \mathbf{y} \rangle$ on a real vector space X is a real-valued mapping of the pair of elements $\mathbf{x}, \mathbf{y} \in X$ with the properties: $\langle \mathbf{x}, \mathbf{y} \rangle = \langle \mathbf{y}, \mathbf{x} \rangle$; $\langle \alpha\mathbf{x}, \mathbf{y} \rangle = \alpha\langle \mathbf{x}, \mathbf{y} \rangle$ for all $\alpha \in \Re$; $\langle \mathbf{x} + \mathbf{y}, \mathbf{z} \rangle = \langle \mathbf{x}, \mathbf{z} \rangle + \langle \mathbf{y}, \mathbf{z} \rangle$ for all $\mathbf{z} \in X$; and $\langle \mathbf{x}, \mathbf{x} \rangle > 0 \ \forall \mathbf{x} \neq \mathbf{0}$. A real vector space with an inner product defined is called a *(real) inner product space*. An inner product space is a normed vector space and hence a metric space with the distance function defined as $d(\mathbf{x},\mathbf{y}) = \|\mathbf{x}-\mathbf{y}\| = \sqrt{\langle \mathbf{x}-\mathbf{y},\mathbf{x}-\mathbf{y}\rangle}$. For any two vectors $\mathbf{x}$ and $\mathbf{y}$ in an inner product space, we have the *Schwarz inequality* in the form $|\langle \mathbf{x},\mathbf{y}\rangle| \leq \|\mathbf{x}\|\|\mathbf{y}\|$. The equality holds if and only if $\mathbf{x}$ and $\mathbf{y}$ are dependent. The vectors $\mathbf{x}$ and $\mathbf{y}$ are *orthogonal* if $\langle \mathbf{x},\mathbf{y}\rangle = 0$. Let $\{\mathbf{x}_i\}$ be a set of elements in an inner product space X. $\{\mathbf{x}_i\}$ is an *orthogonal set* if $\langle \mathbf{x}_i,\mathbf{x}_j\rangle = 0, \forall i \neq j$. If in addition every vector in the set has unit norm, the set is *orthonormal*.

A *Hilbert space* is a complete inner product space with the norm induced by its inner product. For example, $\Re^n$ is a Hilbert space with inner product $\langle \mathbf{x},\mathbf{y}\rangle = \sum x_i y_i$. Suppose scalar functions $x(t)$ and $y(t)$ are defined in a domain D, then L_2 is a Hilbert space with the inner product $\langle x,y\rangle = \int_D x(t)y(t)dt$.

Best Approximation Problem in Hilbert Spaces

Let U be a set of vectors in a Hilbert space H. The *algebraic span $S(U)$* is defined as the set of all finite linear combinations of vectors $\mathbf{x}_i \in U$ (Stakgold and Holst, 2011). The set $\overline{S}(U)$ is the closure of $S(U)$ and is called the *closed span* of U. For example, if $U = \{1,x,x^2,...\} \subset L_2$, then $S(U)$ is the set of all polynomials, whereas $\overline{S}(U) = L_2$. The set U is a *spanning set* of H if $\overline{S}(U)$ is dense in H. The Hilbert space H is *separable* if it contains a countable spanning set U. The space L_2 is separable since the countable set $\{1,x,x^2,...\}$ is a spanning set. Any finite-dimensional Hilbert space is separable because its basis is a countable spanning set. Since an infinite-dimensional space cannot have a finite spanning set, a separable infinite-dimensional Hilbert space must contain a countably infinite set $U=\{\mathbf{x}_i\}$ so that each vector $\mathbf{x} \in H$ can be approximated to any desired accuracy by a linear combination of a finite number of elements of U. This can be rewritten as: given $\varepsilon > 0$ and $\mathbf{x} \in H$, there exist an integer n such that $\left\|\mathbf{x} - \sum c_i \mathbf{x}_i\right\| < \varepsilon$ where

$c_i \in \Re, i = 1,...,n$. It can be prove that a separable Hilbert space H contains an orthonormal spanning set, and the spanning set is necessarily an orthonormal basis of H. An orthonormal basis is also known as a *complete* orthonormal set.

The best approximation problem in a separable Hilbert space is to approximate an arbitrary vector $\mathbf{x} \in H$ by a linear combination of the given independent set $U = \{\mathbf{x}_1,...,\mathbf{x}_n\}$. Since the set of linear combination of $\mathbf{x}_1,...,\mathbf{x}_n$ is an n-dimensional linear manifold M_n, an orthonormal basis $\{\mathbf{e}_1,...,\mathbf{e}_n\}$ for M_n can be constructed from U by using the Gram-Schmidt procedure. Therefore, the *approximation error* can be calculated as

$$\left\| \mathbf{x} - \sum c_i \mathbf{e}_i \right\|^2 = \|\mathbf{x}\|^2 + \sum \left| \langle \mathbf{x}, \mathbf{e}_i \rangle - c_i \right|^2 - \sum \left| \langle \mathbf{x}, \mathbf{e}_i \rangle \right|^2$$

Hence, the minimum error can be obtained when $c_i = \langle \mathbf{x}, \mathbf{e}_i \rangle$. These c_i are known as the *Fourier coefficients* of $\mathbf{x}$ with respect to the orthonormal basis $\{\mathbf{e}_1,...,\mathbf{e}_n\}$. With these coefficients, the vector $\mathbf{x} \in H$ is approximated as $\sum \langle \mathbf{x}, \mathbf{e}_i \rangle \mathbf{e}_i$, and the approximation error becomes $\left\| \mathbf{x} - \sum \langle \mathbf{x}, \mathbf{e}_i \rangle \mathbf{e}_i \right\|^2 = \|\mathbf{x}\|^2 - \sum \left| \langle \mathbf{x}, \mathbf{e}_i \rangle \right|^2$. If an additional vector $\mathbf{e}_{n+1}$ is included into the orthonormal set, the vector $\mathbf{x} \in H$ is thus approximated by the series in a summation of the $n+1$st term, i.e., with the extra term $\langle \mathbf{x}, \mathbf{e}_{i+1} \rangle \mathbf{e}_{i+1}$. This implies that previously calculated Fourier coefficients do not need to be recalculated. It is also seen that the approximation error will get smaller when the orthonormal set is taken larger. Hence, the best approximation to $\mathbf{x}$ improves as we use more terms in the orthonormal set. As the number of terms used goes to infinity, the approximating series becomes the *Fourier series*, and hence we have $\|\mathbf{x}\|^2 = \sum \left| \langle \mathbf{x}, \mathbf{e}_i \rangle \right|^2$ which is known as the *Parseval's identity*. Convergence of the Fourier series can be proved by using the Riesz-Fischer theorem with the fact that the partial sum of the infinite series $\sum \left| \langle \mathbf{x}, \mathbf{e}_i \rangle \right|^2$ is monotonically increasing and is bounded above by $\|\mathbf{x}\|^2$. Consequently, it is easy to prove that $\lim_{i \to \infty} \langle \mathbf{x}, \mathbf{e}_i \rangle = 0$, i.e., the coefficients of the Fourier series vanish as $i \to \infty$.

Orthogonal Functions

In the previous subsection we have reviewed the general framework for the best approximation problem in the Hilbert space. Here, we would like to restrict the scope to the function approximation problem using orthogonal functions.

The set of real-valued functions $\{\phi_i(x)\}$ defined over some interval $[a,b]$ is said to form an *orthogonal set* on that interval if

$$\int_a^b \phi_i(x)\phi_j(x)dx \begin{cases} =0 & i \neq j \\ \neq 0 & i=j \end{cases}$$

An orthogonal set $\{\phi_i(x)\}$ on $[a,b]$ having the property $\int_a^b \phi_i^2(x)dx = 1$ for all i is called an *orthonormal set* on $[a,b]$. The set of real-valued functions $\{\phi_i(x)\}$ defined over some interval $[a,b]$ is orthogonal with respect to the weight function $p(x)$ on that interval if

$$\int_a^b p(x)\phi_i(x)\phi_j(x)dx \begin{cases} =0 & i \neq j \\ \neq 0 & i=j \end{cases}$$

Any set of functions orthogonal with respect to a weigh function $p(x)$ can be converted into a set of functions orthogonal to 1 simply by multiplying each member of the set by $\sqrt{p(x)}$ if $p(x)>0$ on that interval. For any set of orthonormal functions $\{\phi_i(x)\}$ on $[a,b]$, an arbitrary function $f(x)$ can be represented in terms of $\phi_i(x)$ by a series

$$f(x) = c_1\phi_1(x) + c_2\phi_2(x) + \cdots + c_n\phi_n(x) + \cdots$$

This series is called a *generalized Fourier series* of $f(x)$ and its coefficients are Fourier coefficients of $f(x)$ with respect to $\{\phi_i(x)\}$. Multiplying by $\phi_n(x)$ and integrating over the interval $[a,b]$ and using the orthogonality property, the series becomes

$$\int_a^b f(x)\phi_n(x)dx = c_n \int_a^b \phi_n^2(x)dx$$

Hence, the coefficient c_n can be obtained from the quotient

$$c_n = \frac{\int_a^b f(x)\phi_n(x)dx}{\int_a^b \phi_n^2(x)dx}$$

It should be noted that although the orthogonality property can be used to determine all coefficients, it is not sufficient to conclude convergence of the series. To guarantee convergence of the approximating series, the orthogonal set should be complete. An orthogonal set $\{\phi_i(x)\}$ on $[a,b]$ is said to be *complete* if the relation $\int_a^b g(x)\phi_i(x)dx = 0$ can hold for all values of i only if $g(x)$ can have non-zero values in a measure zero set in $[a,b]$. Here, $g(x)$ is called a *null function* on $[a,b]$ satisfying $\int_a^b g^2(x)dx = 0$. It is easy to prove that if $\{\phi_i(x)\}$ is a complete orthonormal set on $[a,b]$ and the expansion $c_1\phi_1(x)+c_2\phi_2(x)+\cdots+c_n\phi_n(x)+\cdots$ of $f(x)$ converges and can be integrated term by term, then the sum of the series differs from $f(x)$ by at most a null function.

Examples of Orthonormal Functions

Since there are many areas of applications of orthonormal functions, a sizable body of literature can be easily found. In this subsection, we consider some of the orthonormal functions that are frequently encountered in engineering problems and useful in our applications.

1. Taylor polynomials

In the calculus courses, it is well known that given a function $f(x)$ and a point c in the domain of f, suppose the function is n-times differentiable at c, then we can construct a polynomial

$$P_n(x) = f(c) + f'(c)(x-c) + \frac{f''(c)}{2!}(x-c)^2 + \ldots + \frac{f^{(n)}(c)}{n!}(x-c)^n$$

where $P_n(x)$ is called the nth-degree Taylor polynomial approximation of f at c. The Taylor polynomial is not well suited to approximate a function $f(x)$ over an interval $[a,b]$ if the approximation is to be uniformly accurate over the entire domain. Taylor polynomial approximation is known to yield very small error near a given point, but the error increases in a considerable amount as we move away from that point. The following orthogonal polynomials, however, can give a more uniform approximation error over the specified interval.

2. Chebyshev polynomials

The set of Chebyshev polynomials is orthogonal with respect to the weight function $(1-x^2)^{-\frac{1}{2}}$ on the interval $[-1,1]$. The first two polynomials are $T_0(x) = 1$ and $T_1(x) = x$, and the remaining polynomials can be determined by the recurrence relation

$$T_{n+1}(x) = 2xT_n(x) - T_{n-1}(x)$$

for all $n = 1,2,...$ For convenience, we list the first 7 polynomials below

$$T_0(x) = 1$$

$$T_1(x) = x$$

$$T_2(x) = 2x^2 - 1$$

$$T_3(x) = 4x^3 - 3x$$

$$T_4(x) = 8x^4 - 8x^2 + 1$$

$$T_5(x) = 16x^5 - 20x^3 + 5x$$

$$T_6(x) = 32x^6 - 48x^4 + 18x^2 + 1$$

$$T_7(x) = 64x^7 - 112x^5 + 56x^3 - 7x$$

3. Legendre polynomials

The set of Legendre polynomials is orthogonal with respect to the weight function $p(x) = 1$ on the interval [-1,1]. The first two polynomials are $L_0(x) = 1$ and $L_1(x) = x$, and the remaining polynomials can be determined by the recurrence relation

$$(n+1)L_{n+1}(x) = (2n+1)xL_n(x) - nL_{n-1}(x)$$

for all $n = 1,2,...$ Here, we list the first 7 polynomials for convenience

$$L_0(x) = 1$$

$$L_1(x) = x$$

$$L_2(x) = \frac{1}{2}(3x^2 - 1)$$

$$L_3(x) = \frac{1}{2}(5x^3 - 3x)$$

$$L_4(x) = \frac{1}{8}(35x^4 - 30x^2 + 3)$$

$$L_5(x) = \frac{1}{8}(63x^5 - 70x^3 + 15x)$$

$$L_6(x) = \frac{1}{16}(231x^6 - 315x^4 + 105x^2 - 5)$$

$$L_7(x) = \frac{1}{16}(429x^7 - 693x^5 + 315x^3 - 35x)$$

4. Hermite polynomials

The set of Hermite polynomials is orthogonal with respect to the weight function $p(x) = e^{-x^2}$ on the interval $(-\infty, \infty)$. The first two polynomials are $H_0(x) = 1$ and $H_1(x) = 2x$, and the remaining polynomials can be determined by the recurrence relation

$$H_{n+1}(x) = 2xH_n(x) - 2nH_{n-1}(x)$$

for all $n = 1,2,...$ Here, we list the first 7 polynomials as

$$H_0(x) = 1$$

$$H_1(x) = 2x$$

$$H_2(x) = 4x^2 - 2$$

$$H_3(x) = 8x^3 - 12x$$

$$H_4(x) = 16x^4 - 48x^2 + 12$$

$$H_5(x) = 32x^5 - 160x^3 + 120x$$

$$H_6(x) = 64x^6 - 480x^4 + 720x^2 - 120$$

$$H_7(x) = 128x^7 - 1344x^5 + 3360x^3 - 1680x$$

5. Laguerre polynomials

The set of Laguerre polynomials is orthogonal with respect to the weight function $p(x) = e^{-x}$ on the interval $[0,\infty)$. The first two polynomials are $L_0(x) = 1$ and $L_1(x) = -x+1$, and the remaining polynomials can be determined by the recurrence relation

$$L_{n+1}(x) = (2n+1-x)L_n(x) - n^2 L_{n-1}(x)$$

for all $n = 1,2,...$ The following are the first 7 polynomials

$$L_0(x) = 1$$

$$L_1(x) = -x+1$$

$$L_2(x) = x^2 - 4x + 2$$

$$L_3(x) = -x^3 + 9x^2 - 18x + 6$$

$$L_4(x) = x^4 - 16x^3 + 72x^2 - 96x + 24$$

$$L_5(x) = -x^5 + 25x^4 - 200x^3 + 600x^2 - 600x + 120$$

$$L_6(x) = x^6 - 36x^5 + 450x^4 - 2400x^3 + 5400x^2 - 4320x + 720$$

$$L_7(x) = -x^7 + 49x^6 - 882x^5 + 7350x^4 - 29400x^3 + 52920x^2 - 35280x + 5040$$

6. Bessel polynomials

The set of Bessel polynomials is orthogonal with respect to the weight function $p(x) = x$ on the interval $[0,b]$ in the form

$$\int_0^b x J_n(k_i x) J_n(k_j x) dx = 0 \qquad \forall i \neq j$$

where k_i, $i=1,2,\ldots$ are real numbers so that $J_n(k_i b) = 0$, i.e., they are distinct roots of $J_n = 0$. These roots for $n=0$, 1 are listed here for reference.

$$J_0(x) = 0 \text{ for } x = 2.405, \quad 5.520, \quad 8.654, \quad 11.792, \quad 14.931$$

$$J_1(x) = 0 \text{ for } x = 0, \quad 3.832, \quad 7.016, \quad 10.173, \quad 13.324$$

The Bessel polynomials can be calculated with

$$J_n(x) = x^n \sum_{m=0}^{\infty} \frac{(-1)^m x^{2m}}{2^{2m+n} m!(n+m)!}$$

In particular, for $n=0$, 1, the Bessel polynomials are

$$J_0(x) = 1 - \frac{x^2}{2^2} + \frac{x^4}{2^2 \cdot 4^2} - \frac{x^6}{2^2 \cdot 4^2 \cdot 6^2} + \cdots$$

$$J_1(x) = \frac{x}{2} - \frac{x^3}{2^2 \cdot 4} + \frac{x^5}{2^2 \cdot 4^2 \cdot 6} - \frac{x^7}{2^2 \cdot 4^2 \cdot 6^2 \cdot 8} + \cdots$$

These two series converge very rapidly, so that they are useful in computations. The recurrence relation below can also be used to find other Bessel polynomials based on $J_0(x)$ and $J_1(x)$ given above.

$$J_{n+1}(x) = -J_{n-1}(x) + \frac{2n}{x} J_n(x)$$

With the orthogonality property, we can represent a given function $f(x)$ in a series of the form in $[0,b]$ with a given n

$$f(x) = \sum_{i=1}^{\infty} c_i J_n(k_i x)$$

This series is called a *Fourier-Bessel series* or simply a *Bessel series*.

7. Fourier series

A bounded periodical function $f(x)$ can be expanded in the form

$$f(x) = \frac{a_0}{2} + \sum_{n=1}^{\infty} \left[a_n \cos \frac{n\pi x}{T} + b_n \sin \frac{n\pi x}{T} \right] \tag{1}$$

if in any one period it has at most a finite number of local extreme values and a finite number of discontinuities. The above representation is called the *Fourier series* of function $f(x)$. The constants a_0, a_n and b_n, $n=1,2,3,\ldots$ are called *Fourier coefficients*, and the value $2T$ is the period of $f(x)$. It can be proved that the Fourier series converges to $f(x)$ at all points where $f(x)$ is continuous and converges to the average of the right- and left-hand limits of $f(x)$ at each point where it is discontinuous.

Function Norms and Normed Function Spaces

A real-valued function defined on $\Re_+$ is *measurable* if and only if it is the limit of a sequence of piecewise constant functions over $\Re^+$

except for some measure zero sets. Let $f(t): \mathfrak{R}_+ \to \mathfrak{R}$ be a measurable function, then its *p*-norm and infinity norm are defined as

$$\|f\|_p = \left[\int_0^\infty |f(t)|^p \, dt \right]^{\frac{1}{p}} \quad p \in [1, \infty)$$

$$\|f\|_\infty = \sup_{t \in [0, \infty)} |f(t)| \quad p = \infty$$

The normed function spaces with $p = 1, 2, \infty$ are defined as

$$L_1 = \left\{ f(t): \mathfrak{R}_+ \to \mathfrak{R} \Big| \|f\|_1 = \int_0^\infty |f(t)| \, dt < \infty \right\}$$

$$L_2 = \left\{ f(t): \mathfrak{R}_+ \to \mathfrak{R} \Big| \|f\|_2 = \sqrt{\int_0^\infty |f(t)|^2 \, dt} < \infty \right\}$$

$$L_\infty = \left\{ f(t): \mathfrak{R}_+ \to \mathfrak{R} \Big| \|f\|_\infty = \sup_{t \in [0, \infty)} |f(t)| < \infty \right\}$$

Let $\mathbf{f}: \mathfrak{R}_+ \to \mathfrak{R}^n$ with $\mathbf{f}(t) = [f_1(t) \cdots f_n(t)]^T$ be a measurable vector function, then the corresponding *p*-norm spaces are defined as

$$L_1^n = \left\{ \mathbf{f}(t): \mathfrak{R}_+ \to \mathfrak{R}^n \Big| \|\mathbf{f}\|_1 = \int_0^\infty \sum_{i=1}^n |f_i(t)| \, dt < \infty \right\}$$

$$L_2^n = \left\{ \mathbf{f}(t): \mathfrak{R}_+ \to \mathfrak{R}^n \Big| \|\mathbf{f}\|_2 = \sqrt{\int_0^\infty \sum_{i=1}^n |f_i(t)|^2 \, dt} < \infty \right\}$$

$$L_\infty^n = \left\{ \mathbf{f}(t): \mathfrak{R}_+ \to \mathfrak{R}^n \Big| \|\mathbf{f}\|_\infty = \max_{1 \le i \le n} |f_i(t)| < \infty \right\}$$

Representations for Approximation

In this subsection some representations for approximation of scalar functions, vectors and matrices using finite-term orthonormal functions are reviews. Let us consider a set of real-valued functions $\{z_i(t)\}$ that are orthonormal in $[t_1, t_2]$ such that

$$\int_{t_1}^{t_2} z_i(t) z_j(t) dt = \begin{cases} 0, & i \neq j \\ 1, & i = j \end{cases}$$

With the definition of the inner product $< f,g >= \int_{t_1}^{t_2} f(t)g(t)dt$ and its corresponding norm $\|f\| = \sqrt{< f,f >}$, the space of functions for which $\|f\|$ exists and is finite is a Hilbert space. If $\{z_i(t)\}$ is an orthonormal basis then every $f(t)$ with $\|f\|$ finite can be expanded in the form

$$f(t) = \sum_{i=1}^{\infty} w_i z_i(t)$$

where $w_i =< f,z_i >$ is the Fourier coefficient, and the series converges in the sense of mean square as

$$\lim_{k \to \infty} \int_{t_1}^{t_2} \left| f(t) - \sum_{i=1}^{k} w_i z_i(t) \right|^2 dt = 0$$

This implies that any function $f(t)$ in the current Hilbert space can be approximated to arbitrarily prescribed accuracy by finite linear combinations of the orthonormal basis $\{z_i(t)\}$ as

$$f(t) \approx \sum_{i=1}^{k} w_i z_i(t)$$

An excellent property of this approximation is its linear parameterization of the time-varying function $f(t)$ into a basis function vector $\mathbf{z}(t) = [z_1(t) \cdots z_k(t)]^T$ and a time-invariant coefficient vector $\mathbf{w} = [w_1 \cdots w_k]^T$, i.e.,

$$f(t) \approx \mathbf{w}^T \mathbf{z}(t)$$

We would like to abuse the notation by writing the approximation as

$$f(t) = \mathbf{w}^T \mathbf{z}(t) \tag{2}$$

provided a sufficient number of the basis functions are used. In this book, the above approximation is used to represent time-varying parameters in the system dynamic equation. The time-varying vector $\mathbf{z}(t)$ is known while $\mathbf{w}$ is an unknown constant vector. With this approximation, the estimation of the unknown time-varying function $f(t)$ is reduced to the estimation of a vector of unknown constants $\mathbf{w}$.

In the following, three representations are introduced for approximating a matrix $\mathbf{M}(t) \in \Re^{p \times q}$. By letting $q=1$, the same technique can be used to approximate vectors.

Representation 1: We may use the approximation in (2) to represent individual matrix elements. Let $\mathbf{w}_{ij}, \mathbf{z}_{ij} \in \Re^k$ for all i, j, then matrix $\mathbf{M}$ is represented to be

$$\mathbf{M} = \begin{bmatrix} m_{11} & m_{12} & \cdots & m_{1q} \\ m_{21} & m_{22} & \cdots & m_{2q} \\ \vdots & \vdots & \ddots & \vdots \\ m_{p1} & m_{p2} & \cdots & m_{pq} \end{bmatrix} = \begin{bmatrix} \mathbf{w}_{11}^T \mathbf{z}_{11} & \mathbf{w}_{12}^T \mathbf{z}_{12} & \cdots & \mathbf{w}_{1q}^T \mathbf{z}_{1q} \\ \mathbf{w}_{21}^T \mathbf{z}_{21} & \mathbf{w}_{22}^T \mathbf{z}_{22} & \cdots & \mathbf{w}_{2q}^T \mathbf{z}_{2q} \\ \vdots & \vdots & \ddots & \vdots \\ \mathbf{w}_{p1}^T \mathbf{z}_{p1} & \mathbf{w}_{p2}^T \mathbf{z}_{p2} & \cdots & \mathbf{w}_{pq}^T \mathbf{z}_{pq} \end{bmatrix}$$

An operation $\otimes$ can be defined to separate the above representation into two parts as

$$\begin{bmatrix} \mathbf{w}_{11}^T \mathbf{z}_{11} & \mathbf{w}_{12}^T \mathbf{z}_{12} & \cdots & \mathbf{w}_{1q}^T \mathbf{z}_{1q} \\ \mathbf{w}_{21}^T \mathbf{z}_{21} & \mathbf{w}_{22}^T \mathbf{z}_{22} & \cdots & \mathbf{w}_{2q}^T \mathbf{z}_{2q} \\ \vdots & \vdots & \ddots & \vdots \\ \mathbf{w}_{p1}^T \mathbf{z}_{p1} & \mathbf{w}_{p2}^T \mathbf{z}_{p2} & \cdots & \mathbf{w}_{pq}^T \mathbf{z}_{pq} \end{bmatrix} = \begin{bmatrix} \mathbf{w}_{11}^T & \mathbf{w}_{12}^T & \cdots & \mathbf{w}_{1q}^T \\ \mathbf{w}_{21}^T & \mathbf{w}_{22}^T & \cdots & \mathbf{w}_{2q}^T \\ \vdots & \vdots & \ddots & \vdots \\ \mathbf{w}_{p1}^T & \mathbf{w}_{p2}^T & \cdots & \mathbf{w}_{pq}^T \end{bmatrix} \otimes \begin{bmatrix} \mathbf{z}_{11} & \mathbf{z}_{12} & \cdots & \mathbf{z}_{1q} \\ \mathbf{z}_{21} & \mathbf{z}_{22} & \cdots & \mathbf{z}_{2q} \\ \vdots & \vdots & \ddots & \vdots \\ \mathbf{z}_{p1} & \mathbf{z}_{p2} & \cdots & \mathbf{z}_{pq} \end{bmatrix}$$

Or, we may write the above relation in the following form

$$\mathbf{M} = \mathbf{W}^T \otimes \mathbf{Z}$$

where $\mathbf{W}$ is a matrix containing all $\mathbf{w}_{ij}$ and $\mathbf{Z}$ is a matrix of all $\mathbf{z}_{ij}$. Since this is not a conventional operation of matrices, dimensions of all involved matrices do not follow the rule for matrix multiplication. Here, $\mathbf{W}^T$ is a $p \times kq$ matrix and $\mathbf{Z}$ is a $kp \times q$ matrix, but the dimension of $\mathbf{M}$

after the operation is still $p \times q$. This notation can be used to facilitate the derivation of update laws.

Representation 2: Let us assume that all matrix elements are approximated using the same number, say β, of orthonormal functions, and then the matrix $\mathbf{M}(t) \in \mathfrak{R}^{p \times q}$ can be represented in the conventional form for matrix multiplications

$$\mathbf{M} = \mathbf{W}^T \mathbf{Z}$$

where $\mathbf{W} \in \mathfrak{R}^{pq\beta \times p}$ and $\mathbf{Z} \in \mathfrak{R}^{pq\beta \times q}$ are in the form

$$\mathbf{W}^T = \begin{bmatrix} \mathbf{w}_{11}^T & \mathbf{0} & \cdots & \mathbf{0} & | & \mathbf{w}_{12}^T & \mathbf{0} & \cdots & \mathbf{0} & | & \cdots & | & \mathbf{w}_{1q}^T & \mathbf{0} & \cdots & \mathbf{0} \\ \mathbf{0} & \mathbf{w}_{21}^T & \cdots & \mathbf{0} & | & \mathbf{0} & \mathbf{w}_{22}^T & \cdots & \mathbf{0} & | & \cdots & | & \mathbf{0} & \mathbf{w}_{2q}^T & \cdots & \mathbf{0} \\ \vdots & \vdots & \ddots & \vdots & | & \vdots & \vdots & \ddots & \vdots & | & \cdots & | & \vdots & \vdots & \ddots & \vdots \\ \mathbf{0} & \mathbf{0} & \cdots & \mathbf{w}_{p1}^T & | & \mathbf{0} & \mathbf{0} & \cdots & \mathbf{w}_{p2}^T & | & \cdots & | & \mathbf{0} & \mathbf{0} & \cdots & \mathbf{w}_{pq}^T \end{bmatrix} \quad (3a)$$

$$\mathbf{Z}^T = \begin{bmatrix} \mathbf{z}_{11}^T & \mathbf{z}_{21}^T & \cdots & \mathbf{z}_{p1}^T & | & \mathbf{0} & \mathbf{0} & \cdots & \mathbf{0} & | & \cdots & | & \mathbf{0} & \mathbf{0} & \cdots & \mathbf{0} \\ \mathbf{0} & \mathbf{0} & \cdots & \mathbf{0} & | & \mathbf{z}_{12}^T & \mathbf{z}_{22}^T & \cdots & \mathbf{z}_{p2}^T & | & \cdots & | & \mathbf{0} & \mathbf{0} & \cdots & \mathbf{0} \\ \vdots & \vdots & \ddots & \vdots & | & \vdots & \vdots & \ddots & \vdots & | & \cdots & | & \vdots & \vdots & \ddots & \vdots \\ \mathbf{0} & \mathbf{0} & \cdots & \mathbf{0} & | & \mathbf{0} & \mathbf{0} & \cdots & \mathbf{0} & | & \cdots & | & \mathbf{z}_{1q}^T & \mathbf{z}_{2q}^T & \cdots & \mathbf{z}_{pq}^T \end{bmatrix} \quad (3b)$$

The matrix elements $\mathbf{w}_{ij}$ and $\mathbf{z}_{ij}$ are $\beta \times 1$ vectors. It can be easily check that

$$\mathbf{M} = \mathbf{W}^T \mathbf{Z} = \begin{bmatrix} \mathbf{w}_{11}^T \mathbf{z}_{11} & \mathbf{w}_{12}^T \mathbf{z}_{12} & \cdots & \mathbf{w}_{1q}^T \mathbf{z}_{1q} \\ \mathbf{w}_{21}^T \mathbf{z}_{21} & \mathbf{w}_{22}^T \mathbf{z}_{22} & \cdots & \mathbf{w}_{2q}^T \mathbf{z}_{2q} \\ \vdots & \vdots & \ddots & \vdots \\ \mathbf{w}_{p1}^T \mathbf{z}_{p1} & \mathbf{w}_{p2}^T \mathbf{z}_{p2} & \cdots & \mathbf{w}_{pq}^T \mathbf{z}_{pq} \end{bmatrix} \quad (4)$$

In this representation, we use the usual matrix operation to represent $\mathbf{M}$, but the sizes of $\mathbf{W}$ and $\mathbf{Z}$ are apparently much larger than those in the representation 1. Since this representation is compatible to all conventional matrix operations, it is used in this book for representing functions, vectors and matrices.

Representation 3: In the above representations, all matrix elements are approximated by the same number of orthonormal functions. In many applications, however, it may be desirable to use different number of orthonormal functions for different matrix elements. Suppose the component form of a vector field $\mathbf{f}(\mathbf{x})$ is written as

$$\mathbf{f}(\mathbf{x}) = [f_1(\mathbf{x}) \quad f_2(\mathbf{x}) \quad \cdots \quad f_m(\mathbf{x})]^T$$

and we may approximate the real-valued function $f_i(\mathbf{x})$, $i=1,\ldots,m$ as

$$f_i(\mathbf{x}) = \mathbf{w}_{f_i}^T \mathbf{z}_{f_i}$$

where $\mathbf{w}_{f_i}, \mathbf{z}_{f_i} \in \Re^{p_i \times 1}$ and p_i is the number of terms of the basis functions selected to approximate f_i. Hence, we may further have

$$\mathbf{f}(\mathbf{x}) = [\mathbf{w}_{f_1}^T \mathbf{z}_{f_1} \quad \mathbf{w}_{f_2}^T \mathbf{z}_{f_2} \quad \cdots \quad \mathbf{w}_{f_i}^T \mathbf{z}_{f_m}]^T$$

$$= \begin{bmatrix} w_{11}z_{11} + w_{12}z_{12} + \cdots + w_{1p_1}z_{1p_1} \\ w_{21}z_{21} + w_{22}z_{22} + \cdots + w_{2p_2}z_{2p_2} \\ \vdots \\ w_{m1}z_{m1} + w_{m2}z_{m2} + \cdots + w_{mp_m}z_{mp_m} \end{bmatrix}$$

Define $p_{\max} = \max_{i=1,\cdots,m} p_i$ and let $w_{ij} = 0$ for all $i=1,\ldots,m$ and $j > p_i$, then we may further write

$$\mathbf{f}(\mathbf{x}) = \begin{bmatrix} w_{11} & 0 & \cdots & 0 \\ 0 & w_{21} & \cdots & 0 \\ \vdots & \vdots & \ddots & \vdots \\ 0 & 0 & \cdots & w_{m1} \end{bmatrix} \begin{bmatrix} z_{11} \\ z_{21} \\ \vdots \\ z_{m1} \end{bmatrix} + \cdots + \begin{bmatrix} w_{1p_{\max}} & 0 & \cdots & 0 \\ 0 & w_{2p_{\max}} & \cdots & 0 \\ \vdots & \vdots & \ddots & \vdots \\ 0 & 0 & \cdots & w_{mp_{\max}} \end{bmatrix} \begin{bmatrix} z_{1p_{\max}} \\ z_{2p_{\max}} \\ \vdots \\ z_{mp_{\max}} \end{bmatrix}$$

Define

$$\mathbf{W}_i = \begin{bmatrix} w_{1i} & 0 & \cdots & 0 \\ 0 & w_{2i} & \cdots & 0 \\ \vdots & \vdots & \ddots & \vdots \\ 0 & 0 & \cdots & w_{mi} \end{bmatrix}, \quad \mathbf{z}_i = [z_{1i} \quad z_{2i} \quad \cdots \quad z_{mi}]^T$$

where $i=1,...,p_{max}$, and then we finally have

$$\mathbf{f}(\mathbf{x}) = \sum_{i=1}^{p_{max}} \mathbf{W}_i \mathbf{z}_i$$

For approximating the matrix $\mathbf{M}(t) \in \mathfrak{R}^{p \times q}$, we may rewrite it into a row vector as $\mathbf{M} = [\mathbf{m}_1 \cdots \mathbf{m}_q]$ where $\mathbf{m}_i \in \mathfrak{R}^p$. Therefore, we may approximate $\mathbf{m}_i$ using the technique above as

$$\mathbf{M} = \begin{bmatrix} \sum_{i=1}^{p_{1max}} \mathbf{W}_{1i} \mathbf{z}_{1i} & \cdots & \sum_{i=1}^{p_{qmax}} \mathbf{W}_{qi} \mathbf{z}_{qi} \end{bmatrix}$$

2.2 Lyapunov Stability Theory

The Lyapunov stability theory is widely used in the analysis and design of control systems. To ensure closed loop stability and boundedness of internal signals, all controllers derived in this book will be based on rigorous mathematical proofs via Lyapunov or Lyapunov-like theories. The concept of stability in the sense of Lyapunov will be introduced first in this section followed by Lyapunov stability theorems for autonomous and non-autonomous systems. The invariant set theorem will be reviewed to facilitate the proof for asymptotically stability of autonomous systems when only negative semi-definite of the time derivative of the Lyapunov function can be concluded. In addition, a Lyapunov-like technique summarized in Barbalat's lemma will also be reviewed.

Concepts of Stability

Let us consider a nonlinear dynamic system described by the differential equation

$$\dot{\mathbf{x}} = \mathbf{f}(\mathbf{x}, t) \tag{5}$$

where $\mathbf{x} \in \Re^n$ and $\mathbf{f} : \Re^n \times \Re_+ \to \Re^n$. If the function $\mathbf{f}$ dose not explicitly depend on time t, i.e., the system is in the form

$$\dot{\mathbf{x}} = \mathbf{f}(\mathbf{x}) \tag{6}$$

then it is called an *autonomous system*; otherwise, a *non-autonomous system*. The state trajectory of an autonomous system is independent of the initial time, but that of a non-autonomous system is generally not. Therefore, in studying the behavior of a non-autonomous system, we have to consider the initial time explicitly. The system (5) is *linear* if $\mathbf{f}(\mathbf{x}, t) = \mathbf{A}(t)\mathbf{x}$ for some mapping $\mathbf{A}(\cdot) : \Re_+ \to \Re^{n \times n}$. If the matrix A is a function of time, the system is *linear time-varying*; otherwise, *linear time-invariant*.

Stability is the most important property of a control system. The concept of stability of a dynamic system is usually related to the ability to remain in a state regardless of small perturbations. This leads to the definition of the concept of the equilibrium state or equilibrium point. A state $\mathbf{x}_e$ is said to be an *equilibrium point* of (5), if $\mathbf{f}(\mathbf{x}_e, t) = \mathbf{0}$ for all $t > 0$. For simplicity, we often transform the system equations in such a way that the equilibrium point is the origin of the state space.

The equilibrium point $\mathbf{x}_e = \mathbf{0}$ of the autonomous system (6) is said to be (i) *stable*, if $\forall R > 0, \exists r > 0$ such that $\|\mathbf{x}(0)\| < r \Rightarrow \|\mathbf{x}(t)\| < R$, $\forall t \geq 0$; (ii) *asymptotically stable*, if it is stable and if $\exists r_1 > 0$ such that $\|\mathbf{x}(0)\| < r_1$ implies that $\mathbf{x}(t) \to 0$ as $t \to \infty$; (iii) *exponentially stable*, if $\exists \alpha, \lambda > 0$, such that $\|\mathbf{x}(t)\| \leq \alpha \|\mathbf{x}(0)\| e^{-\lambda t}$ for all $t > 0$ in some neighborhood of the origin; (iv) *globally asymptotically (or exponentially) stable*, if the property holds for any initial condition.

The equilibrium point $\mathbf{x}_e = \mathbf{0}$ of the non-autonomous system (5) is said to be (i) *stable at* t_0, if $\forall R > 0, \exists r(R, t_0) > 0$ such that $\|\mathbf{x}(t_0)\| < r(R, t_0) \Rightarrow \|\mathbf{x}(t)\| < R$, $\forall t \geq t_0$; otherwise the equilibrium point is *unstable* (ii) *asymptotically stable at* t_0 if it is stable and $\exists r_1(t_0) > 0$ such that $\|\mathbf{x}(t_0)\| < r_1(t_0)$ implies that $\|\mathbf{x}(t)\| \to 0$ as $t \to \infty$; (iii) *uniformly stable* if $\forall R > 0, \exists r(R) > 0$ such that $\|\mathbf{x}(t_0)\| < r(R) \Rightarrow \|\mathbf{x}(t)\| < R$, $\forall t \geq t_0$; (iv) *uniformly asymptotically stable* if it is uniformly stable and $\exists r_1 > 0$ such that $\|\mathbf{x}(t_0)\| < r_1$ implies that $\|\mathbf{x}(t)\| \to 0$ uniformly as $t \to \infty$; (v) *exponentially stable*, if $\exists \alpha, \lambda > 0$, such that $\|\mathbf{x}(t)\| \leq \alpha \|\mathbf{x}(t_0)\| e^{-\lambda(t-t_0)}$ for all $t > t_0$ in some ball around the origin; (vi) *globally* (*uniformly*) *asymptotically* (*or exponentially*) *stable*, if the property holds for any initial conditions. It is noted that exponential stability always implies uniform asymptotic stability. Likewise, uniform asymptotic stability always implies asymptotic stability, but the converse is not generally true.

A continuous function $\alpha(r) : \Re \to \Re$ is said to belong to *class K* if (i) $\alpha(0) = 0$; (ii) $\alpha(r) > 0 \ \forall r > 0$; (iii) $\alpha(r_1) \geq \alpha(r_2) \ \forall r_1 > r_2$. A continuous function $V(\mathbf{x}, t) : \Re^n \times \Re_+ \to \Re$ is *locally positive definite* if there exists a class K function $\alpha(\cdot)$ such that $V(\mathbf{x}, t) \geq \alpha(\|\mathbf{x}\|)$ for all $t \geq 0$ in the neighborhood N of the origin of $\Re^n$. It is *positive definite* if $N = \Re^n$. A continuous function $V(\mathbf{x}, t) : \Re^n \times \Re_+ \to \Re$ is *locally decrescent* if there exists a class K function $\beta(\cdot)$ such that $V(\mathbf{x}, t) \leq \beta(\|\mathbf{x}\|)$ for all $t \geq 0$ in the neighborhood N of the origin of $\Re^n$. It is *decrescent* if $N = \Re^n$. A continuous function $V(\mathbf{x}, t) : \Re^n \times \Re_+ \to \Re$ is *radially unbounded* if $V(\mathbf{x}, t) \to \infty$ uniformly in time as $\|\mathbf{x}\| \to \infty$. If function $V(\mathbf{x}, t)$ is locally positive definite and has continuous partial derivatives, and if its time derivative along the trajectory of (5) is negative semi-definite then it is called a *Lyapunov function* for (5).

Lyapunov Stability Theorem for Autonomous Systems

Given the autonomous system (6) with an equilibrium point at the origin, and let N be a neighborhood of the origin, then the origin is (i) *stable* if there exists a scalar function $V(\mathbf{x}) > 0$　$\forall \mathbf{x} \in N$ such that $\dot{V}(\mathbf{x}) \leq 0$; (ii) *asymptotic stable* if $V(\mathbf{x}) > 0$ and $\dot{V}(\mathbf{x}) < 0$; (iii) *globally asymptotically stable* if $V(\mathbf{x}) > 0$, $\dot{V}(\mathbf{x}) < 0$ and $V(\mathbf{x})$ is radially unbounded.

LaSalle's Theorem (Invariant Set Theorem)

Given the autonomous system (6) suppose $V(\mathbf{x}) > 0$ and $\dot{V}(\mathbf{x}) \leq 0$ along the system trajectory. Then (6) is asymptotically stable if $\dot{V}$ does not vanish identically along any trajectory of (6) other than the trivial solution $\mathbf{x} = \mathbf{0}$. The result is global if the properties hold for the entire state space and $V(\mathbf{x})$ is radially unbounded.

Lyapunov Stability Theorem for Non-autonomous Systems

Given the non-autonomous system (5) with an equilibrium point at the origin, and let N be a neighborhood of the origin, then the origin is (i) *stable at* t_0 if $\forall \mathbf{x} \in N$, there exists a scalar function $V(\mathbf{x},t)$ such that $V(\mathbf{x},t) > 0$ and $\dot{V}(\mathbf{x},t) \leq 0$; (ii) *uniformly stable* if $V(\mathbf{x},t) > 0$ and decrescent and $\dot{V}(\mathbf{x},t) \leq 0$; (iii) *asymptotically stable at* t_0 if $V(\mathbf{x},t) > 0$ and $\dot{V}(\mathbf{x},t) < 0$; (iv) *globally asymptotically stable* if $\forall \mathbf{x} \in \mathfrak{R}^n$, there exists a scalar function $V(\mathbf{x},t)$ such that $V(\mathbf{x},t) > 0$ and $\dot{V}(\mathbf{x},t) < 0$ and $V(\mathbf{x},t)$ is radially unbounded; (v) *uniformly asymptotically stable* if $\forall \mathbf{x} \in N$, there exists a scalar function $V(\mathbf{x},t)$ such that $V(\mathbf{x},t) > 0$ and decrescent and $\dot{V}(\mathbf{x},t) < 0$; (vi) *globally uniformly asymptotically stable* if $\forall \mathbf{x} \in \mathfrak{R}^n$, there exists a scalar function $V(\mathbf{x},t)$ such that $V(\mathbf{x},t) > 0$ and decrescent and is radially unbounded and $\dot{V}(\mathbf{x},t) < 0$; (vii) *exponentially stable* if there exits $\alpha, \beta, \gamma > 0$ such that $\forall \mathbf{x} \in N$, $\alpha \|\mathbf{x}\|^2 \leq V(\mathbf{x},t) \leq \beta \|\mathbf{x}\|^2$ and $\dot{V}(\mathbf{x},t) \leq -\gamma \|\mathbf{x}\|^2$; (viii) *globally exponentially stable* if it is exponentially stable and $V(\mathbf{x},t)$ is radially unbounded.

Lyapunov-like Analysis: Barbalat's Lemma

La Salle's theorem is very useful in the stability analysis of autonomous systems when asymptotic stability is desired but only with negative semi-definite result for the time derivative of the Lyapunov function. Unfortunately, La Salle's theorem does not apply to non-autonomous systems. Therefore, to conclude asymptotic stability of a non-autonomous system with $\dot{V} \leq 0$, we need to find a new approach. A simple and powerful tool called Barbalat's lemma can be used to partially remedy this situation. Let $f(t)$ be a differentiable function, then Barbalat's lemma states that if $\lim_{t \to \infty} f(t) = k < \infty$ and $\dot{f}(t)$ is uniformly continuous, then $\lim_{t \to \infty} \dot{f}(t) = 0$. It can be proved that a differentiable function is uniformly continuous if its derivative is bounded. Hence, the lemma can be rewritten as: if $\lim_{t \to \infty} f(t) = k < \infty$ and $\ddot{f}(t)$ exists and is bounded, then $\dot{f} \to 0$ as $t \to \infty$. In the Lyapunov stability analysis, Barbalat's lemma can be applied in the fashion similar to La Salle's theorem: If $V(\mathbf{x}, t)$ is lower bounded, $\dot{V} \leq 0$, and $\ddot{V}$ is bounded, then $\dot{V} \to 0$ as $t \to \infty$. It should be noted that the Lyapunov function is only required to be lower bounded in stead of positive definite. In addition, we can only conclude convergence of $\dot{V}$, not the states. In this book, we would like to use the other form of Barbalat's lemma to prove closed loop stability. If we can prove that a time function e is bounded and square integrable, and its time derivative is also bounded, then e is going to converge to zero asymptotically. It can be restated as: if $e \in L_\infty \cap L_2$ and $\dot{e} \in L_\infty$, then $e \to 0$ as $t \to \infty$.

2.3 Frobenius Theorem

In chapter 3 we are going to find a coordinate transformation to decouple the underactuated dynamics, and a set of partial differential equations is to be solved. The Frobenius theorem provides a simple way to determine the solvability of this set of partial differential equations without actually solve them. In this section, we review various background materials of the Frobenius theorem from differential geometry.

A *diffeomorphism* is a differentiable function with a differentiable inverse. A *differentiable manifold* is a topological space which is locally diffeomorphic to the Euclidean space. In this book, we would like to call a differentiable manifold as a *manifold* for short and regard it as a subset in $\Re^n$ defined by the p scalar functions

$$h_1(x_1,...,x_n) = 0$$
$$h_2(x_1,...,x_n) = 0$$
$$\vdots$$
$$h_p(x_1,...,x_n) = 0 \qquad p < n$$

This set of scalar functions can be collected as a vector-valued function $\mathbf{h} : \Re^n \rightarrow \Re^p$ satisfying the relationship

$$\mathbf{h}(\mathbf{x}) = \mathbf{0}$$

where $\mathbf{x} = [x_1 \quad \cdots \quad x_n]^T \in \Re^n$ and $\mathbf{h}(\mathbf{x}) = [h_1(\mathbf{x}) \quad \cdots \quad h_p(\mathbf{x})]^T \in \Re^p$. Assume that the differentials of $h_i(\mathbf{x})$, $i = 1,...,p$ are linearly independent at all points on the manifold, then we may say that it has the dimension $m = n - p$. Hence, for each point $\mathbf{x}$ on an m-dimensional manifold M, there is a *tangent space* $T_x M$ satisfying all conditions for a vector space. A *smooth vector field* on an m-dimensional manifold M is an infinitely differentiable function $\mathbf{f} : M \rightarrow T_x M$ which can also be written as an m-dimensional column vector

$$\mathbf{f}(\mathbf{x}) = \begin{bmatrix} f_1(\mathbf{x}) \\ \vdots \\ f_m(\mathbf{x}) \end{bmatrix} \in \Re^m$$

Therefore, for a single-input autonomous control system in the form

$$\dot{\mathbf{x}} = \mathbf{f}(\mathbf{x}) + \mathbf{g}(\mathbf{x})u$$

we may regard $\mathbf{f}(\mathbf{x}) \in \mathfrak{R}^n$ and $\mathbf{g}(\mathbf{x}) \in \mathfrak{R}^n$ as two smooth vector fields on an n-dimensional manifold M. The Jacobian of the smooth vector field $\mathbf{f}(\mathbf{x}) \in \mathfrak{R}^n$ is defined as

$$\frac{\partial \mathbf{f}}{\partial \mathbf{x}} = \begin{bmatrix} \dfrac{\partial f_1}{\partial x_1} & \cdots & \dfrac{\partial f_1}{\partial x_n} \\ \vdots & \ddots & \vdots \\ \dfrac{\partial f_n}{\partial x_1} & \cdots & \dfrac{\partial f_n}{\partial x_n} \end{bmatrix} \in \mathfrak{R}^{n \times n}$$

The *Lie bracket* of smooth vector fields $\mathbf{f}(\mathbf{x}) \in \mathfrak{R}^n$ and $\mathbf{g}(\mathbf{x}) \in \mathfrak{R}^n$ is a vector field defined as

$$[\mathbf{f}, \mathbf{g}] = \frac{\partial \mathbf{g}}{\partial \mathbf{x}} \mathbf{f} - \frac{\partial \mathbf{f}}{\partial \mathbf{x}} \mathbf{g} \tag{7}$$

Let $\mathbf{f}_1(\mathbf{x}), \ldots, \mathbf{f}_q(\mathbf{x})$ be smooth vector fields on a manifold M that are linearly independent at all points on M. Then, a *distribution* Δ is defined as the linear span represented as

$$\Delta = sp\{\mathbf{f}_1(\mathbf{x}) \quad \ldots \quad \mathbf{f}_q(\mathbf{x})\}$$

which is therefore a q-dimensional subspace of the m-dimensional tangent space $T_\mathbf{x} M$. A distribution $\Delta = sp\{\mathbf{f}_1(\mathbf{x}) \quad \ldots \quad \mathbf{f}_q(\mathbf{x})\}$ on $\mathfrak{R}^n$ is *involutive* if and only if there are scalar functions $\alpha_{ijk} : \mathfrak{R}^n \to \mathfrak{R}$ such that

$$[\mathbf{f}_i(\mathbf{x}), \mathbf{f}_j(\mathbf{x})] = \sum_{k=1}^{q} \alpha_{ijk} \mathbf{f}_k(\mathbf{x}) \tag{8}$$

for all i, j and k.

When finding coordinate transformations for decoupling underactuated dynamics, we will encounter the solvability problem of a set of first-order partial differential equations. We are now ready to state the necessary and sufficient condition for the existence of the solution for these partial differential equations. Let us firstly look at a simple example of the solvability for the set of two partial differential equations

$$\frac{\partial h}{\partial x_1} f_{11}(\mathbf{x}) + \frac{\partial h}{\partial x_2} f_{12}(\mathbf{x}) + \frac{\partial h}{\partial x_3} f_{13}(\mathbf{x}) = 0$$

$$\frac{\partial h}{\partial x_1} f_{21}(\mathbf{x}) + \frac{\partial h}{\partial x_2} f_{22}(\mathbf{x}) + \frac{\partial h}{\partial x_3} f_{23}(\mathbf{x}) = 0$$

where $f_{ij}, i = 1,2$ and $j = 1,2,3$ are known smooth functions of $\mathbf{x} = [x_1 \quad x_2 \quad x_3]^T \in \mathfrak{R}^3$, while $h(\mathbf{x})$ is an unknown scalar function to be determined. Let us define the *gradient* of $h(\mathbf{x})$ as a row vector

$$\frac{\partial h(\mathbf{x})}{\partial \mathbf{x}} = \begin{bmatrix} \frac{\partial h}{\partial x_1} & \frac{\partial h}{\partial x_2} & \frac{\partial h}{\partial x_3} \end{bmatrix} \in \mathfrak{R}^{1 \times 3}$$

then the set of differential equation can be rewritten as

$$\frac{\partial h}{\partial \mathbf{x}} \mathbf{f}_1(\mathbf{x}) = 0$$

$$\frac{\partial h}{\partial \mathbf{x}} \mathbf{f}_2(\mathbf{x}) = 0$$

where $\mathbf{f}_i(\mathbf{x}) = [f_{i1}(\mathbf{x}) \quad f_{i2}(\mathbf{x}) \quad f_{i3}(\mathbf{x})]^T$, $i = 1,2$ are smooth vector fields. Hence, the set of partial differential equations is uniquely defined by $\mathbf{f}_1(\mathbf{x})$ and $\mathbf{f}_2(\mathbf{x})$. If a solution $h(\mathbf{x})$ exists, then we say that the set of vector fields $\{\mathbf{f}_1(\mathbf{x}), \mathbf{f}_2(\mathbf{x})\}$ is completely integrable. In general cases, the distribution $\Delta = sp\{\mathbf{f}_1(\mathbf{x}) \quad ... \quad \mathbf{f}_m(\mathbf{x})\}$ on $\mathfrak{R}^n$ is said to be *completely integrable* if and only if there are $n - m$ linearly independent scalar functions $h_1(\mathbf{x}), h_2(\mathbf{x}), ..., h_{n-m}(\mathbf{x})$ satisfying the system of partial differential equations

$$\frac{\partial h_i}{\partial \mathbf{x}} \mathbf{f}_j(\mathbf{x}) = 0, \quad 1 \le i \le n - m, \quad 1 \le j \le m \tag{9}$$

The Frobenius theorem states that a distribution is completely integrable if and only if it is involutive. In other words, the set of partial differential equations in (9) is solvable if an only if the distribution $\Delta = sp\{\mathbf{f}_1(\mathbf{x}) \quad \dots \quad \mathbf{f}_m(\mathbf{x})\}$ is involutive.

2.4 Sliding Control

A practical control system should be designed to ensure system stability and performance to be invariant under perturbations from internal parameter variation, unmodeled dynamics excitation and external disturbances. For nonlinear systems, the sliding control is perhaps the most popular approach to achieve the robust performance requirement. In the sliding control, a sliding surface is designed so that the system trajectory is forced to converge to the surface by some worst-case control efforts. Once on the surface, the system dynamics is reduced to a stable linear time invariant system which is irrelevant to the perturbations no matter from internal or external sources. Convergence of the output error is then easily achieved. In this section, we are going to review the sliding controller design including two smoothing techniques to eliminate the chattering activity in the control effort.

Let us consider a non-autonomous system

$$x^{(n)} = f(\mathbf{x},t) + g(\mathbf{x},t)u + d(t) \tag{10}$$

where $\mathbf{x} = [x \quad \dot{x} \quad \cdots \quad x^{(n-1)}]^T \in \mathfrak{R}^n$ is the state vector, $x \in \mathfrak{R}$ the output of interest, and $u \in \mathfrak{R}$ the control input. The function $f(\mathbf{x},t) \in \mathfrak{R}$ and the disturbance $d(t) \in \mathfrak{R}$ are both unknown functions of time, but bounds of their variations should be available. The control gain function $g(\mathbf{x},t) \in \mathfrak{R}$ is assumed to be non-singular for all admissible $\mathbf{x}$ and for all time t. In the following derivation, we would like to design a sliding controller with the knowledge of $g(\mathbf{x},t)$ first, and then a controller is constructed with unknown $g(\mathbf{x},t)$. Let us assume that $f(\mathbf{x},t)$ and $d(t)$ can be modeled as additive uncertainties

$$f = f_m + \Delta f \tag{11a}$$

$$d = d_m + \Delta d \tag{11b}$$

where f_m and d_m are known nominal values of f and d, respectively. The uncertain terms Δf and Δd are assumed to be bounded by some known functions $\alpha(\mathbf{x},t) > 0$ and $\beta(\mathbf{x},t) > 0$, respectively, as

$$|\Delta f| \leq \alpha(\mathbf{x},t) \tag{12a}$$

$$|\Delta d| \leq \beta(\mathbf{x},t) \tag{12b}$$

Since the system contains uncertainties, the inversion-based controller

$$u = \frac{1}{g(\mathbf{x},t)}[-f(\mathbf{x},t) - d(t) + v] \tag{13}$$

is not realizable, where v is some terms to complete the desired dynamics. We would like to design a tracking controller so that the output x tracks the desired trajectory x_d asymptotically regardless of the presence of uncertainties. Let us define a *sliding surface* $s(\mathbf{x},t) = 0$ as a desired error dynamics, where $s(\mathbf{x},t)$ is a linear stable differential operator acting on the tracking error $e = x - x_d$ as

$$s = \left(\frac{d}{dt} + \lambda\right)^{n-1} e \tag{14}$$

where $\lambda > 0$ determines the behavior of the error dynamics. Selection of the sliding surface is not unique, but the one in (14) is preferable simply because it is linear and it will result in a relative degree one dynamics, i.e., u appears when we differentiate s once. One way to achieve output error convergence is to find a control u such that the state trajectory converges to the sliding surface. Once on the surface, the system behaves like a stable linear system

$$\left(\frac{d}{dt} + \lambda\right)^{n-1} e = 0$$

therefore, asymptotic convergence of the tracking error can be obtained. Now, the problem is how to drive the system trajectory to the sliding surface. With $s(\mathbf{x}, t) = 0$ as the boundary, the state space can be decomposed into two parts: the one with $s > 0$ and the other with $s < 0$. Intuitively, to make the sliding surface attractive, we can design a control u so that s will decrease in the $s > 0$ region, and it will increase in the $s < 0$ region. This condition is called the *sliding condition* which can be written in a compact form

$$s\dot{s} < 0 \tag{15}$$

Using (14), the sliding surface for system (10) is of the form

$$s = c_1 e + c_2 \dot{e} + \cdots + c_{n-1} e^{(n-2)} + c_n e^{(n-1)} \tag{16}$$

where $c_k = \dfrac{(n-1)! \lambda^{n-k}}{(n-k)!(k-1)!}$, $k = 1, \ldots, n-1$, and $c_n = 1$. Let us differentiate s with respect to time once to make u appear

$$
\begin{aligned}
\dot{s} &= c_1 \dot{e} + \cdots + c_{n-1} e^{(n-1)} + c_n e^{(n)} \\
&= c_1 \dot{e} + \cdots + c_{n-1} e^{(n-1)} + x^{(n)} - x_d^{(n)} \\
&= c_1 \dot{e} + \cdots + c_{n-1} e^{(n-1)} + f + gu + d - x_d^{(n)}
\end{aligned}
\tag{17}
$$

Since g is known and non-singular, we may select the control u as

$$u = \frac{1}{g}[-c_1 \dot{e} - \cdots - c_{n-1} e^{(n-1)} - f_m - d_m + x_d^{(n)} - \eta_1 \operatorname{sgn}(s)] \tag{18}$$

where $\eta_1 > 0$ is a design parameter to be determined, and $\operatorname{sgn}(s)$ is the signum function defined as

$$\operatorname{sgn}(s) = \begin{cases} 1 & \text{if } s > 0 \\ -1 & \text{if } s < 0 \end{cases}$$

so that (17) becomes

$$\dot{s} = \Delta f + \Delta d - \eta_1 \operatorname{sgn}(s) \tag{19}$$

To satisfy the sliding condition (15), let us multiply both sides of (19) with s as

$$s\dot{s} = (\Delta f + \Delta d)s - \eta_1|s|$$
$$\le (\alpha + \beta)|s| - \eta_1|s| \tag{20}$$

By picking $\eta_1 = \alpha + \beta + \eta$ with $\eta > 0$, the above inequality becomes

$$s\dot{s} \le -\eta|s| \tag{21}$$

Therefore, with the controller (18), the sliding surface (16) is attractive, and the tracking error converges asymptotically regardless of the uncertainties in f and d, once the sliding surface is reached. Now, let us consider the case when $g(\mathbf{x},t)$ is not available, but we do know that it is non-singular for all admissible state and time t, and its variation bound is known with $0 < g_{\min} \le g \le g_{\max}$. Instead of the additive uncertainty model we used previously for f and d, a multiplicative model is chosen for g as

$$g = g_m \Delta g \tag{22}$$

where $g_m = \sqrt{g_{\min} g_{\max}}$ is the nominal value of g and Δg satisfies the relation

$$0 \le \gamma_{\min} \equiv \frac{g_{\min}}{g_m} \le \Delta g \le \frac{g_{\max}}{g_m} \equiv \gamma_{\max} \tag{23}$$

In this case, the controller is chosen as

$$u = \frac{1}{g_m}[-c_1\dot{e} - \cdots - c_{n-1}e^{(n-1)} - f_m - d_m + x_d^{(n)} - \eta_1 \operatorname{sgn}(s)] \tag{24}$$

Substituting (24) into (17), we have

$$\dot{s} = (1 - \Delta g)(c_1\dot{e} + \cdots + c_{n-1}e^{(n-1)} + f_m + d_m - x_d^{(n)}) + \Delta f + \Delta d - \Delta g \eta_1 \operatorname{sgn}(s) \tag{25}$$

Multiplying both sides with s, equation (25) becomes

$$s\dot{s} = (1 - \Delta g)(c_1\dot{e} + \cdots + c_{n-1}e^{(n-1)} + f_m + d_m - x_d^{(n)})s + (\Delta f + \Delta d)s - \Delta g \eta_1|s|$$
$$\le (1 - \gamma_{\min})|c_1\dot{e} + \cdots + c_{n-1}e^{(n-1)} + f_m + d_m - x_d^{(n)}||s| + (\alpha + \beta)|s| - \gamma_{\min}\eta_1|s| \tag{26}$$

The parameter η_1 can thus be selected as

$$\eta_1 = \frac{1}{\gamma_{min}}[(1-\gamma_{min})|c_1\dot{e}+...+c_{n-1}e^{(n-1)}+f_m+d_m-x_d^{(n)}|+(\alpha+\beta)+\eta]\ (27)$$

where η is a positive number. Therefore, we can also have the result in (21).

Smoothed Sliding Control Law

Both controllers (18) and (24) contain the switching function sgn(s). In practical implementation, the switching induced from this function will sometimes result in high frequency control activities called chattering. Consequently, the tracking performance degrades, and the high-frequency unmodeled dynamics may be excited. In some cases, the switching controller has to be modified with a continuous approximation. One approach is to use the saturation function sat(σ) defined below instead of the signum function sgn(s)

$$\text{sat}(\sigma) = \begin{cases} \sigma & \text{if } |\sigma| \le \phi \\ \text{sgn}(\sigma) & \text{if } |\sigma| > \phi \end{cases} \tag{28}$$

where $|s| \le \phi$ is called the *boundary layer* of the sliding surface and $\phi > 0$ is the thickness of the boundary layer. When s is outside the boundary layer, i.e., $|s| > \phi$, the sliding controller with sgn(s) is exactly the same as the one with $sat\left(\dfrac{s}{\phi}\right)$. Hence, the boundary layer is also attractive. When s is inside the boundary layer, equation (19) becomes

$$\dot{s} + \eta_1 \frac{s}{\phi} = \Delta f + \Delta d \tag{29}$$

This implies that the signal s is the output of a stable first-order filter whose input is the bounded model error $\Delta f + \Delta d$. Thus, the chattering behavior can indeed be eliminated with proper selection of the filer bandwidth and as long as the high-frequency unmodeled dynamics is not excited. One drawback of this smoothed sliding controller is the degradation of the tracking accuracy. At best we can say that once the signal s converges to the boundary layer, the output tracking error is bounded by the value ϕ.

Instead of the saturation function, we may also use $\dfrac{s}{\phi}$ to replace the signum function to have a smoothed version of the sliding controller. This selection is very easy in implementation, because the robust term is linear in the signal s. For example, controller (18) can be smoothed in the form

$$u = \frac{1}{g}\left[-c_1 \dot{e} - \cdots - c_{n-1} e^{(n-1)} - f_m - d_m + x_d^{(n)} - \eta_1 \frac{s}{\phi} \right] \qquad (30)$$

To justify its effectiveness, the following analysis is performed. When s is outside the boundary layer, (19) can be rewritten in the form

$$\dot{s} = \Delta f + \Delta d - \eta_1 \frac{s}{\phi} \qquad (31)$$

With the selection of $\eta_1 = \alpha + \beta + \eta$, the sliding condition can be checked as

$$s\dot{s} = (\Delta f + \Delta d)s - \eta_1 \frac{s^2}{\phi}$$

$$\leq (\alpha + \beta)|s| - (\alpha + \beta + \eta)\frac{s^2}{\phi} \qquad (32)$$

$$= (\alpha + \beta)|s|\left[1 - \frac{|s|}{\phi} \right] - \eta \frac{s^2}{\phi}$$

Since $|s| > \phi$, i.e., when outside the boundary layer, we may have the result $s\dot{s} \leq -\eta \dfrac{s^2}{\phi}$. Hence, the boundary layer is still attractive. When s is inside the boundary layer, equation (29) can be obtained; therefore, effective chattering elimination can be achieved. Let us now consider the case when $g(\mathbf{x},t)$ is unknown and the sliding controller is smoothed with $\dfrac{s}{\phi}$ as

$$u = \frac{1}{g_m}\left[\bar{u} + x_d^{(n)} - \eta_1\frac{s}{\phi}\right]$$

(33)

where $\bar{u} = -c_1\dot{e} - \cdots - c_{n-1}e^{(n-1)} - f_m - d_m$. By selecting η_1 according to (27), the sliding condition is checked with

$$s\dot{s} \leq [(1-\beta_{\min})|\bar{u}| + \alpha + \beta]|s|\left[1 - \frac{|s|}{\phi}\right] - \eta\frac{s^2}{\phi}$$

(34)

If s is outside the boundary layer, (34) implies $s\dot{s} \leq -\eta\dfrac{s^2}{\phi}$, i.e., all trajectories will eventually converge to the boundary layer even though the system contains uncertainties. Since inside the boundary layer there is also an equivalent first order filter dynamics, the chattering activity can be effectively eliminated. The output tracking error, however, can only be concluded to be uniformly bounded. There is one more drawback for the smoothing using $\dfrac{s}{\phi}$, i.e., the initial control effort may become enormously large if there is a significant difference between the desired trajectory and the initial state. To overcome this problem, desired trajectory and initial conditions should be carefully selected.

2.5 Adaptive Control

Adaptive control and robust control are two main approaches for controlling systems containing uncertainties and disturbances. The sliding control introduced in the previous section is one of the robust designs widely used in the literature. In this section, the well-known MRAC (Model Reference Adaptive Control) is reviewed as an example in the traditional adaptive approach. For an adaptive controller to be feasible the system structure is assumed to be known and a set of unknown constant system parameters (or equivalently the corresponding controller parameters) are to be estimated so that the closed loop stability is ensured via a certainty equivalence based controller. In this section, the MRAC for a linear time-invariant scalar system is introduced first, followed by the investigation of the persistent excitation condition for the convergence of estimated parameters.

MRAC of LTI Scalar Systems

Consider a linear time-invariant system described by the differential equation

$$\dot{x}_p = a_p x_p + b_p u \tag{35}$$

where $x_p \in \Re$ is the state of the plant and $u \in \Re$ the control input. The parameters a_p and b_p are unknown constants, but $\mathrm{sgn}(b_p)$ is available. The pair (a_p, b_p) is controllable. The problem is to design a control u and an update law so that all signals in the closed loop are bounded and the system output x_p tracks the output x_m of the reference model

$$\dot{x}_m = a_m x_m + b_m r \tag{36}$$

asymptotically, where a_m and b_m are known constants with $a_m < 0$, and r is a bounded reference signal. If plant parameters a_p and b_p are available, the model reference control (MRC) rule can be designed as

$$u = a x_p + b r \tag{37}$$

where $a = \dfrac{a_m - a_p}{b_p}$ and $b = \dfrac{b_m}{b_p}$ are perfect gains for transforming dynamics in (35) into (36). Since the values of a_p and b_p are not given, we may not select these perfect gains to complete the MRC design in (37) and the model reference adaptive control rule is constructed instead

$$u = \hat{a}(t) x_p + \hat{b}(t) r \tag{38}$$

where $\hat{a}$ and $\hat{b}$ are estimates of a and b, respectively, and proper update laws are to be selected to give $\hat{a} \to a$ and $\hat{b} \to b$. Define the output tracking error as

$$e = x_p - x_m \tag{39}$$

then the error dynamics can be computed as following

$$\dot{e} = a_m e + b_p(\hat{a} - a)x_p + b_p(\hat{b} - b)r \tag{40}$$

Define the parameter errors $\tilde{a} = \hat{a} - a$ and $\tilde{b} = \hat{b} - b$, then equation (40) is further written as

$$\dot{e} = a_m e + b_p \tilde{a} x_p + b_p \tilde{b} r \tag{41}$$

This is the dynamics of the output error e, which is a stable linear system driven by the parameter errors. Therefore, if update laws are found to have convergence of these parameter errors, convergence of the output error follows. To find these update laws for $\hat{a}$ and $\hat{b}$, let us define a Lyapunov function candidate

$$V(e, \tilde{a}, \tilde{b}) = \frac{1}{2} e^2 + \frac{1}{2}|b_p|(\tilde{a}^2 + \tilde{b}^2) \tag{42}$$

Taking the time derivative of V along the trajectory of (41), we have

$$\begin{aligned}
\dot{V} &= e\dot{e} + |b_p|(\tilde{a}\dot{\tilde{a}} + \tilde{b}\dot{\tilde{b}}) \\
&= e(a_m e + b_p \tilde{a} x_p + b_p \tilde{b} r) + |b_p|(\tilde{a}\dot{\tilde{a}} + \tilde{b}\dot{\tilde{b}}) \\
&= a_m e^2 + \tilde{a}|b_p|[\mathrm{sgn}(b_p)ex_p + \dot{\tilde{a}}] + \tilde{b}|b_p|[\mathrm{sgn}(b_p)er + \dot{\tilde{b}}]
\end{aligned} \tag{43}$$

If the update laws are selected as

$$\dot{\hat{a}} = -\mathrm{sgn}(b_p)ex_p \tag{44a}$$

$$\dot{\hat{b}} = -\mathrm{sgn}(b_p)er \tag{44b}$$

then (43) becomes

$$\dot{V} = a_m e^2 \leq 0 \tag{45}$$

This implies that $e, \tilde{a}, \tilde{b} \in L_\infty$. From the simple derivation

$$\int_0^\infty e^2 dt = -a_m^{-1} \int_0^\infty \dot{V} dt = a_m^{-1}(V_0 - V_\infty) < \infty$$

we know that $e \in L_2$. The result $\dot{e} \in L_\infty$ can easily be concluded from (41). Therefore, it follows from Barbalat's lemma that the output error $e(t)$ converges to zero asymptotically. In summary, the controller (38) together with update laws in (44) make the system (35) track the reference model (36) asymptotically with boundedness of all internal signals. It can be observed in (44) that the update laws are driven by the tracking error e. Once e gets close to zero, the estimated parameters converges to some values. We cannot predict the exact values these parameters will converge to from the above derivation. Let us consider the situation when the system gets into the steady state, i.e., when $t \rightarrow \infty$. The error dynamics (41) becomes

$$\tilde{a} x_p + \tilde{b} r = 0 \tag{46}$$

If r is a constant, then x_p in (46) can be found as $x_p = x_m = kr$, where $k = -\dfrac{b_m}{a_m}$ is the d.c. gain of the reference model (36). Equation (46) further implies

$$k\tilde{a} + \tilde{b} = 0 \tag{47}$$

which is exactly a straight line in the parameter error space. Therefore, for a constant reference input r, both the estimated parameters do not necessarily converge to zero. To investigate the problem of parameter convergence, we need the concept of persistent excitation. A signal $v \in \mathfrak{R}^n$ is said to satisfy the *persistent excitation* (PE) condition if $\exists \alpha, T > 0$ such that

$$\int_{t}^{t+T} \mathbf{v}\mathbf{v}^T dt \geq \alpha \mathbf{I} \tag{48}$$

Define $\mathbf{v} = [x_p \quad r]^T$ and $\tilde{\boldsymbol{\theta}} = [\tilde{a} \quad \tilde{b}]^T$, and equation (46) is able to be represented into the vector form

$$\mathbf{v}^T \tilde{\boldsymbol{\theta}} = [x_p \quad r]\begin{bmatrix} \tilde{a} \\ \tilde{b} \end{bmatrix} = 0 \tag{49}$$

Since $\mathbf{v}\mathbf{v}^T \tilde{\boldsymbol{\theta}} = \mathbf{0}$, its integration in $[t, t+T]$ is

$$\int_{t}^{t+T} \mathbf{v}\mathbf{v}^T \tilde{\boldsymbol{\theta}} dt = \mathbf{0} \tag{50}$$

When $t \to \infty$, update laws in (44) imply $\dot{\tilde{\boldsymbol{\theta}}} \to \mathbf{0}$; therefore, (50) becomes

$$\int_{t}^{t+T} \mathbf{v}\mathbf{v}^T dt \tilde{\boldsymbol{\theta}} = \mathbf{0} \tag{51}$$

Hence, if $\mathbf{v}$ is PE, equation (51) implies $\tilde{\boldsymbol{\theta}} = \mathbf{0}$, i.e., parameter convergence when $t \to \infty$.

2.6 Robust Adaptive Control

The adaptive controller presented in the previous section is developed for LTI systems without external disturbances or unmodeled dynamics. For practical control systems, uncertain parameters may vary with time and the system may contain some non-parametric uncertainties. Rohrs et al. (1985) showed that in the presence of a small amount of measurement noise and high-frequency unmodeled dynamics, an adaptive control system presents slow parameter drift behavior and the system output suddenly diverges sharply after a finite interval of time. For practical implementation, an adaptive control system should be designed to withstand all kinds of non-parametric uncertainties. Some modifications of the adaptive laws have been developed to deal with these problems. In the following, a technique called dead-zone is introduced followed by the review of the well-known σ-modification.

Dead-Zone

Consider the uncertain linear time-invariant system

$$\dot{x}_p = a_p x_p + b_p u + d(t) \tag{52}$$

where a_p is unknown but $b_p \neq 0$ is available. The disturbance $d(t)$ is assumed to be bounded by some $\delta > 0$. A reference model is designed as

$$\dot{x}_m = a_m x_m + b_m r \tag{53}$$

where $a_m < 0$ and $b_m = b_p$. Let $a = \dfrac{a_m - a_p}{b_p}$ and $b = 1$ be ideal feedback gains for the MRAC law (38). Since a_p is not given, a practical feedback law is designed to be

$$u = \hat{a}(t) x_p + br \tag{54}$$

where $\hat{a}(t)$ is the estimate of a. Let $e = x_p - x_m$ and $\tilde{a} = \hat{a} - a$, then the error dynamics is computed as

$$\dot{e} = a_m e + b_p \tilde{a} x_p + d(t) \tag{55}$$

The time derivative of the Lyapunov function

$$V(e, \tilde{a}) = \frac{1}{2}(e^2 + \tilde{a}^2)$$

along the trajectory of (55) is

$$\dot{V} = a_m e^2 + \tilde{a}(b_p e x_p + \dot{\hat{a}}) + ed \tag{56}$$

With the selection of the update law

$$\dot{\hat{a}} = -b_p e x_p \tag{57}$$

(56) becomes

$$\dot{V} = a_m e^2 + ed$$
$$= (d - |a_m|e)e \tag{58}$$
$$\leq (\delta - |a_m||e|)|e|$$

If $\delta - |a_m||e| < 0$, i.e., $|e| > \dfrac{\delta}{|a_m|}$, then (58) implies that V is non-increasing.

Let D be the set where V will grow unbounded, i.e.

$D = \left\{ (e, \tilde{a}) \middle| |e| \leq \dfrac{\delta}{|a_m|} \right\}$; therefore, the modified update law

$$\dot{\hat{a}} = \begin{cases} -b_p e x_p & \text{if } (e, \tilde{a}) \in D^c \\ 0 & \text{if } (e, \tilde{a}) \in D \end{cases} \tag{59}$$

assures boundedness of all signals in the system. The notation D^c denotes the complement of D. The modified update law (59) implies that when the error e is within the dead-zone D, the update law is inactive to avoid possible parameter drift. It should be noted that, however, the asymptotic convergence of the error signal e is no longer valid after the dead-zone modification even when the disturbance is removed.

σ-modification

In applying the dead-zone modification, the upper bound of the disturbance signal is required to be given. Here, a technique called σ-modification is introduced which does not need the information of disturbance bounds.

Instead of (59), the update law (57) is modified as

$$\dot{\hat{a}} = -b_p e x_p - \sigma \hat{a} \tag{60}$$

where σ is a small positive constant. Then (58) becomes

$$\dot{V} = a_m e^2 - \sigma \tilde{a} \hat{a} + ed$$
$$\leq -|a_m|e^2 - \sigma \tilde{a}^2 - \sigma \tilde{a} a + |e|\delta \tag{61}$$

for some unknown $\delta > 0$. Rewrite the two terms in (61) involving e as

$$-|a_m|e^2 + |e|\delta = -\frac{1}{2}\left[\sqrt{|a_m|}|e| - \frac{\delta}{\sqrt{|a_m|}}\right]^2 - \frac{1}{2}|a_m|e^2 + \frac{1}{2}\frac{\delta^2}{|a_m|}$$

$$\leq -\frac{1}{2}|a_m|e^2 + \frac{1}{2}\frac{\delta^2}{|a_m|} \tag{62}$$

Likewise, the rest two terms in (61) are derived as

$$-\sigma\tilde{a}^2 - \sigma\tilde{a}a \leq -\sigma\tilde{a}^2 + \sigma|\tilde{a}||a|$$

$$\leq -\frac{1}{2}\sigma\tilde{a}^2 + \frac{1}{2}\sigma|a|^2 \tag{63}$$

Substituting (62) and (63) into (61), we have

$$\dot{V} \leq -\frac{1}{2}|a_m|e^2 + \frac{1}{2}\frac{\delta^2}{|a_m|} - \frac{1}{2}\sigma\tilde{a}^2 + \frac{1}{2}\sigma|a|^2 \tag{64}$$

Adding and subtracting αV for some $\alpha > 0$, (64) becomes

$$\dot{V} \leq -\alpha V + \frac{1}{2}\frac{\delta^2}{|a_m|} + \frac{1}{2}\sigma|a|^2 + \frac{1}{2}(\alpha - |a_m|)e^2 + \frac{1}{2}(\alpha - \sigma)\tilde{a}^2 \tag{65}$$

Picking $\alpha < \min\{|a_m|, \sigma\}$, we obtain

$$\dot{V} \leq -\alpha V + \frac{1}{2}\frac{\delta^2}{|a_m|} + \frac{1}{2}\sigma|a|^2 \tag{66}$$

Therefore, $\dot{V} \leq 0$, if

$$V \geq \frac{1}{2\alpha}\frac{\delta^2}{|a_m|} + \frac{1}{2\alpha}\sigma|a|^2 \tag{67}$$

This implies that signals in the closed loop system are uniformly bounded. Hence, the additional term $\sigma\hat{a}$ in the update law makes the adaptive control system robust to bounded external disturbances, although bounds of these disturbances are not given. One way to look at the modification is to rewrite (60) as

$$\dot{\hat{a}} = \underbrace{-\sigma \hat{a}}_{\text{stable linear filter}} - b_p e x_p$$

It is seen that the estimated parameter can be regarded as the output of a stable linear filter driven by the input signal $-b_p e x_p$. One drawback of this method is that the origin of the system (55) and (60) is no longer an equilibrium point, i.e., the error signal e will not converge to zero even when the disturbance is removed.

2.7 General Uncertainties

We have seen that, to derive a sliding controller, the variation bounds of the parametric uncertainties should be given. Availability for the knowledge of the uncertainty variation bounds is a must for almost all robust control strategies. This is because the robust controllers need to cover system uncertainties even for the worst case. On the other hand, we also know that for the adaptive controller to be feasible the unknown parameters should be time-invariant. This is also almost true for most adaptive control schemes. Let us now consider the case when a system contains time-varying uncertainties whose variation bounds are not known. Since it is time-varying, traditional adaptive design is not feasible. Because the variation bounds are not given, the robust strategies fail. We would like to call this kind of uncertainties the *general uncertainties*. It is challenging to design controllers for systems containing general uncertainties.

In the following, we are going to have some investigation on the difficulties for the design of adaptive controllers when the system has time-varying parameters, and we will look at the problem in designing robust controllers for systems containing uncertain parameters without knowing their bounds.

MRAC of LTV Systems

In the conventional design of adaptive control systems such as the one introduced in Section 2.5, there is a common assumption that the unknown parameters to be updated should be time-invariant. This can be understood by considering the scalar linear time-varying system

$$\dot{x}_p = a_p(t)x_p + b_p u \tag{68}$$

where a_p is a time-varying unknown parameter and $b_p > 0$ is known. A controller is to be constructed such that the system behaves like the dynamics of the reference model

$$\dot{x}_m = a_m x_m + b_m r \tag{69}$$

where $a_m < 0$ and $b_m = b_p$. Let $a(t) = \dfrac{a_m - a_p(t)}{b_p}$ and $b = 1$ be ideal feedback gains for the MRC law $u = a(t)x_p + br$. Since $a_p(t)$ is not given, a practical feedback law based on MRAC is designed

$$u = \hat{a}(t)x_p + br \tag{70}$$

where $\hat{a}(t)$ is an adjustable parameter of the controller. Let $e = x_p - x_m$ and $\tilde{a}(t) = \hat{a}(t) - a(t)$, then the error dynamics is computed to be

$$\dot{e} = a_m e + b_p \tilde{a} x_p \tag{71}$$

Take the time derivative of the Lyapunov function candidate

$$V(e, \tilde{a}) = \frac{1}{2}e^2 + \frac{1}{2}b_p \tilde{a}^2 \tag{72}$$

along the trajectory of (71), we have

$$\dot{V} = a_m e^2 + b_p \tilde{a}(ex_p + \dot{\hat{a}} - \dot{a}) \tag{73}$$

If we choose the update law

$$\dot{\hat{a}} = -ex_p \tag{74}$$

then (73) becomes

$$\dot{V} = a_m e^2 - b_p \tilde{a} \dot{a} \tag{75}$$

Since both $\tilde{a}$ and $\dot{a}$ are not available, the definiteness of $\dot{V}$ cannot be determined. Therefore, we are not able to conclude anything about the properties of the signals in the closed loop system. From here, we know that the assumption for the unknown parameters to be time-invariant is very important for the feasibility of the design of adaptive controllers. It is equivalent to say that the traditional MRAC fails in controlling systems with time-varying uncertainties.

Sliding Control for Systems with Unknown Variation Bounds

In this subsection, we would like to give an example to show that the design of a robust controller needs to have the knowledge of the variation bound of the uncertain parameters.

Consider a first order uncertain nonlinear system

$$\dot{x} = f(x,t) + g(x,t)u \tag{76}$$

where $f(x,t)$ is a bounded uncertainty and $g(x,t)$ is a known nonsingular function. The uncertainty $f(x,t)$ is modeled as the summation of the known nominal value f_m and the unknown variation Δf .

$$f(x,t) = f_m + \Delta f \tag{77}$$

Since Δf is a bounded function with unknown bounds, there is a positive constant α which is not available satisfying

$$|\Delta f| \le \alpha \tag{78}$$

Let us select the sliding variable $s = x - x_d$, where x_d is the desired trajectory. The dynamics of the sliding variable is computed as

$$\dot{s} = f + gu - x_d \tag{79}$$

By selecting the sliding control law as

$$u = \frac{1}{g}[-f_m + x_d - \eta_1 \operatorname{sgn}(s)] \tag{80}$$

equation (79) becomes

$$\dot{s} = \Delta f - \eta_1 \operatorname{sgn}(s) \tag{81}$$

Multiplying s to the both sides to have

$$\begin{aligned} s\dot{s} &= \Delta f s - \eta_1|s| \\ &\leq (\alpha - \eta_1)|s| \end{aligned} \tag{82}$$

Since α is not given, we may not select η_1. Therefore, the sliding condition cannot be satisfied and the sliding control fails in this case.

Bound Estimation for Uncertain Parameter

An intuitive attempt to circumvent the difficulty encountered in the previous subsection is to estimate α by using conventional adaptive strategies. Since α is a constant, it might be possible to design a proper update law $\dot{\hat{\alpha}}$ for its estimate $\hat{\alpha}$.

Let η be a positive number, then we may pick $\eta_1 = \hat{\alpha} + \eta$ so that equation (82) becomes

$$s\dot{s} \leq \tilde{\alpha}|s| - \eta|s| \tag{83}$$

where $\tilde{\alpha} = \alpha - \hat{\alpha}$. Consider a Lyapunov function candidate

$$V = \frac{1}{2}s^2 + \frac{1}{2}\tilde{\alpha}^2 \tag{84}$$

Its time derivative can be found as

$$\begin{aligned} \dot{V} &= s\dot{s} - \tilde{\alpha}\dot{\hat{\alpha}} \\ &\leq -\eta|s| + \tilde{\alpha}(|s| - \dot{\hat{\alpha}}) \end{aligned} \tag{85}$$

By selecting the update law as

$$\dot{\hat{\alpha}} = |s| \tag{86}$$

equation (85) becomes

$$\dot{V} \le -\eta |s| \tag{87}$$

It seems that the estimation of the uncertainty bound can result in closed loop stability. However, in practical applications, the error signal s will never be zero, and the update law (86) implies an unbounded $\hat{\alpha}$. Therefore, the concept in estimation of the uncertainty bound is not realizable.

2.8 FAT-Based Adaptive Controller Design

In practical realization of control systems, the mathematical model inevitably contains uncertainties. If the variation bounds of these uncertainties are available, traditional robust control strategies such as the Lyapunov redesign and sliding control are applicable. If their bounds are not given, but we know that these uncertainties are time-invariant, various adaptive control schemes are useful. It is possible that system uncertainties are time-varying without knowing their bounds (general uncertainties); therefore, the above tools are not feasible. In this book, we would like to use the function approximation techniques based designs to overcome the given problem. The FAT has been well-known for hundreds of years and there have been wide applications not only in academic fields but also industrial implementations. Therefore, the integration of the control theory with the reliable FAT deserves our attention. The basic idea of the FAT is to represent the general uncertainties by using a set of known basis functions weighted by a set of unknown coefficients (Huang and Kuo, 2001; Huang and Chen, 2004b; Chen and Huang, 2004). Since these coefficients are constants, the Lyapunov designs can thus be applied to derive proper update laws to ensure closed loop stability. This approach has been successfully applied to the control of many systems, such as friction compensation (Alamir,

2002), robot manipulators (Chien and Huang, 2004; Huang et al., 2006; Huang and Chien, 2010; Azlan and Yamaura, 2012, 2013; Kai and Huang, 2013a; Talaei et al., 2013; Li et al., 2013; Kai and Huang, 2014a; Al-Shuka et al., 2014), flexible joint robots (Huang and Chen, 2004a; Chien and Huang, 2007, 2009, 2010a, 2011), jet engine control (Tyan and Lee, 2005), active vehicle suspensions (Chen and Huang, 2005a, 2005b, 2006), flexible link robots (Huang and Liao, 2006), vibration control (Chang and Shaw, 2007), pneumatic servo (Tsai and Huang, 2008a, 2008b), DC motors (Liang et al., 2008; Cong et al., 2009), visual servo (Chien and Huang, 2010b), belt-driven servo (Lee and Huang, 2011), vibration absorbers (Kai and Huang, 2013), and active linearization (Kai and Huang, 2013).

We know that the traditional MRAC is unable to give proper performance to LTV systems. In the first part of this section, we would like to present the FAT-based MRAC for LTV systems without considering the approximation error. The asymptotic convergence can be obtained if a sufficient number of basis functions are used. In the second part of this section, we will investigate the effect of the approximation error in detail. By considering the approximation error in the adaptive loop, the output error can be proved to be uniformly ultimately bounded. The bound for the transient response of the output error can also be estimated as a weighted exponential function plus some constant offset.

FAT-based MRAC for LTV Systems

Let us consider the linear time-varying system (68) again

$$\dot{x}_p = a_p(t)x_p + b_p u \tag{88}$$

We have proved in Section 2.7 that traditional MRAC is infeasible to give stable closed loop system due to the fact that a_p is time-varying. Let us apply the MRAC rule in (70) once again so that the error dynamics becomes

$$\dot{e} = a_m e + b_p(a - \hat{a})x_p \tag{89}$$

where $a(t) = \dfrac{a_m - a_p(t)}{b_p}$ is the perfect gain in the MRAC rule. Since it is time-varying, traditional MRAC design will end up with (75), and no conclusions for closed loop system stability can be obtained. Here, let us represent a and $\hat{a}$ using function approximation techniques shown in (2)

$$
\begin{aligned}
a &= \mathbf{w}^T \mathbf{z} + \varepsilon \\
\hat{a} &= \hat{\mathbf{w}}^T \mathbf{z}
\end{aligned}
\tag{90}
$$

where $\mathbf{w} \in \mathfrak{R}^{n_a}$ is a vector of weightings, $\hat{\mathbf{w}} \in \mathfrak{R}^{n_a}$ is its estimate, $\mathbf{z} \in \mathfrak{R}^{n_a}$ is a vector of basis functions and ε is the approximation error. The positive integer n_a is the number of terms we selected to perform the function approximation. In this case, we would like to assume that sufficient terms are employed so that the approximation error ε is ignorable. Later in this section, we are going to investigate the effect of the approximation error in detail. Define $\tilde{\mathbf{w}} = \mathbf{w} - \hat{\mathbf{w}}$, and then equation (89) can be represented into the form

$$
\dot{e} = a_m e + b_p \tilde{\mathbf{w}}^T \mathbf{z} x_p
\tag{91}
$$

A new Lyapunov-like function candidate is given as

$$
V(e, \tilde{\mathbf{w}}) = \frac{1}{2} e^2 + \frac{1}{2} b_p \tilde{\mathbf{w}}^T \tilde{\mathbf{w}}
\tag{92}
$$

Its time derivative along the trajectory of (91) is computed to be

$$
\begin{aligned}
\dot{V} &= e\dot{e} - b_p \tilde{\mathbf{w}}^T \dot{\hat{\mathbf{w}}} \\
&= a_m e^2 + b_p \tilde{\mathbf{w}}^T (\mathbf{z} x_p e - \dot{\hat{\mathbf{w}}})
\end{aligned}
\tag{93}
$$

By selecting the update law

$$
\dot{\hat{\mathbf{w}}} = \mathbf{z} x_p e
\tag{94}
$$

we may then have

$$
\dot{V} = a_m e^2 \leq 0
\tag{95}
$$

This implies that both e and $\tilde{\mathbf{w}}$ are uniformly bounded. The output error e can also be concluded to be square integrable from (95). In addition, the boundedness of $\dot{e}$ can easily be observed from (91). Hence, it follows from Barbalat's lemma that e will converge to zero asymptotically. It is seen that a simple application of the FAT can effectively solve the problem of MRAC for linear time-varying systems. Since the FAT is also valid for a much wider class of functions, adaptive controls for nonlinear systems with complex dynamics subject to various uncertainties and disturbances are also possible. This is why we can see so many applications of FAT-base adaptive controllers developed in the above mentioned references.

Consideration of Approximation Error

In the previous subsection, we derived the FAT-based MRAC for LTV systems which is not feasible by using traditional MRAC rule. Let us now consider a more general n-th order non-autonomous system in the standard form

$$
\begin{aligned}
\dot{x}_1 &= x_2 \\
\dot{x}_2 &= x_3 \\
&\vdots \\
\dot{x}_{n-1} &= x_n \\
\dot{x}_n &= f(\mathbf{x},t) + g(\mathbf{x},t)u
\end{aligned}
\tag{96}
$$

where $\mathbf{x} = [x_1 \quad x_2 \quad \cdots \quad x_n]^T \in \Omega$, and Ω is a compact subset of $\mathfrak{R}^n$. $f(\mathbf{x},t)$ is an unknown time-varying function with unknown variation bound. The uncertain function $g(\mathbf{x},t)$ is assumed to be bounded by $0 < g_{\min}(\mathbf{x},t) \le g(\mathbf{x},t) \le g_{\max}(\mathbf{x},t)$ for some known functions $g_{\min}$ and $g_{\max}$ for all $\mathbf{x} \in \Omega$ and $t \in [t_0, \infty)$. Let $g_m = \sqrt{g_{\min} g_{\max}}$ be the nominal function of g, and then we may represent g in the form $g = g_m(\mathbf{x},t)\Delta g(\mathbf{x},t)$ where Δg is the multiplicative uncertainty satisfying

$$
0 < \delta_{\min} \equiv \frac{g_{\min}}{g_m} \le \Delta g \le \frac{g_{\max}}{g_m} \equiv \delta_{\max}
$$

The positive numbers δ_{min} and δ_{max} thus defined are with known values. We would like to design a controller such that the system state vector $\mathbf{x}$ tracks the desired trajectory $\mathbf{x}_d \in \Omega_d$, where Ω_d is a compact subset of Ω. Define the tracking error vector as $\mathbf{e} = \mathbf{x} - \mathbf{x}_d = [x_1 - x_{1d} \quad x_2 - x_{2d} \quad \cdots \quad x_n - x_{nd}]^T$. The control law can then be selected as

$$u = \frac{1}{g_m}(-\hat{f} + v - u_r) \tag{97}$$

where $\hat{f}$ is an estimate of f, u_r is a robust term to cover the uncertainties in g, and $v = \dot{x}_{nd} - \sum_{i=0}^{n-1} k_i e_{i+1}$ is to complete the desired dynamics. The coefficients k_i are selected so that the matrix

$$\mathbf{A} = \begin{bmatrix} 0 & 1 & 0 & \cdots & 0 \\ 0 & 0 & 1 & \cdots & 0 \\ \vdots & \vdots & \vdots & \ddots & \vdots \\ 0 & 0 & 0 & \cdots & 1 \\ -k_0 & -k_1 & -k_2 & \cdots & -k_{n-1} \end{bmatrix} \in \Re^{n \times n}$$

is Hurwitz. With the controller (97), the last line of (96) becomes

$$\dot{x}_n = f + \frac{g}{g_m}(-\hat{f} + v - u_r)$$

$$= (f - \hat{f}) + (1 - \Delta g)(\hat{f} - v) + v - \Delta g u_r$$

Equation (96) can thus be written into the form

$$\dot{e}_n + \sum_{i=0}^{n-1} k_i e_{i+1} = (f - \hat{f}) + (1 - \Delta g)(\hat{f} - v) - \Delta g u_r$$

Its state space representation is then

$$\dot{\mathbf{e}} = \mathbf{A}\mathbf{e} + \mathbf{b}[(f - \hat{f}) + (1 - \Delta g)(\hat{f} - v) - \Delta g u_r] \tag{98}$$

where $\mathbf{b} = [0 \quad 0 \quad \cdots \quad 1]^T \in \mathfrak{R}^n$. Since f is a general uncertainty, we may not use traditional adaptive strategies to have stable closed loop system. Let us apply the function approximation techniques to represent f and its estimate as

$$f = \mathbf{w}^T \mathbf{z} + \varepsilon$$
$$\hat{f} = \hat{\mathbf{w}}^T \mathbf{z}$$

Then (98) becomes

$$\dot{\mathbf{e}} = \mathbf{A}\mathbf{e} + \mathbf{b}[\tilde{\mathbf{w}}^T \mathbf{z} + \varepsilon + (1 - \Delta g)(\hat{f} - v) - \Delta g u_r] \qquad (99)$$

where $\tilde{\mathbf{w}} = \mathbf{w} - \hat{\mathbf{w}}$. To find the update law, let us consider the Lyapunov-like function candidate

$$V = \mathbf{e}^T \mathbf{P}\mathbf{e} + \tilde{\mathbf{w}}^T \mathbf{\Gamma} \tilde{\mathbf{w}} \qquad (100)$$

where $\mathbf{P}$ and $\mathbf{\Gamma}$ are positive definite matrices. In addition, $\mathbf{P}$ satisfies the Lyapunov equation $\mathbf{A}^T \mathbf{P} + \mathbf{P}\mathbf{A} = -\mathbf{Q}$ where $\mathbf{Q}$ is some positive definite matrix. Taking the time derivative of (100) along the trajectory of (99), we have

$$\dot{V} = \mathbf{e}^T (\mathbf{A}^T \mathbf{P} + \mathbf{P}\mathbf{A})\mathbf{e} + 2[(1 - \Delta g)(\hat{f} - v) - \Delta g u_r]\mathbf{b}^T \mathbf{P}\mathbf{e}$$
$$+ 2\varepsilon \mathbf{b}^T \mathbf{P}\mathbf{e} + 2\tilde{\mathbf{w}}^T (\mathbf{z}\mathbf{b}^T \mathbf{P}\mathbf{e} - \mathbf{\Gamma}\dot{\hat{\mathbf{w}}})$$
$$\leq -\mathbf{e}^T \mathbf{Q}\mathbf{e} + 2(1 + \delta_{\max})\left|\hat{f} - v\right|\left\|\mathbf{b}^T \mathbf{P}\mathbf{e}\right\| - 2\delta_{\min} u_r \mathbf{b}^T \mathbf{P}\mathbf{e}$$
$$+ 2\varepsilon \mathbf{b}^T \mathbf{P}\mathbf{e} + 2\tilde{\mathbf{w}}^T (\mathbf{z}\mathbf{b}^T \mathbf{P}\mathbf{e} - \mathbf{\Gamma}\dot{\hat{\mathbf{w}}})$$

We may thus select

$$u_r = \frac{1 + \delta_{\max}}{\delta_{\min}}\left|\hat{f} - v\right|\mathrm{sgn}(\mathbf{b}^T \mathbf{P}\mathbf{e}) \qquad (101a)$$

$$\dot{\hat{\mathbf{w}}} = \mathbf{\Gamma}^{-1}(\mathbf{z}\mathbf{b}^T \mathbf{P}\mathbf{e} - \sigma \hat{\mathbf{w}}), \quad \sigma > 0 \qquad (101b)$$

The signum function in (101a) might induce chattering control activity which would excite un-modeled system dynamics. Some modifications can be used to smooth out the control law. The most intuitive way is to replace the signum function with the saturation function as

$$u_r = \frac{1 + \delta_{max}}{\delta_{min}} \left| \hat{f} - v \right| \text{sat}(\mathbf{b}^T \mathbf{Pe}) \tag{102}$$

One drawback for this modification is the reduction in the output tracking accuracy. It is also noted that the σ-modification term in (101b) is to robustify the adaptive loop. With (101), the time derivative of V becomes

$$\begin{aligned}
\dot{V} &\leq -\mathbf{e}^T \mathbf{Qe} + 2\varepsilon \mathbf{b}^T \mathbf{Pe} + 2\sigma \tilde{\mathbf{w}}^T \hat{\mathbf{w}} \\
&= -\mathbf{e}^T \mathbf{Qe} + 2\varepsilon \mathbf{b}^T \mathbf{Pe} + 2\sigma \tilde{\mathbf{w}}^T (\mathbf{w} - \tilde{\mathbf{w}}) \\
&\leq \underbrace{-\lambda_{min}(\mathbf{Q}) \|\mathbf{e}\|^2 + 2\lambda_{max}(\mathbf{P})|\varepsilon|\|\mathbf{e}\|}_{(a)} + 2\sigma \underbrace{[\tilde{\mathbf{w}}^T \mathbf{w} - \|\tilde{\mathbf{w}}\|^2]}_{(b)}
\end{aligned} \tag{103}$$

Let us derive part (a) in (103) using straightforward manipulations as

$$\begin{aligned}
-\lambda_{min}(\mathbf{Q}) \|\mathbf{e}\|^2 + 2\lambda_{max}(\mathbf{P})|\varepsilon|\|\mathbf{e}\| &= -\frac{1}{2} \left(\sqrt{\lambda_{min}(\mathbf{Q})} \|\mathbf{e}\| - \frac{2\lambda_{max}(\mathbf{P})|\varepsilon|}{\sqrt{\lambda_{min}(\mathbf{Q})}} \right)^2 \\
&\quad -\frac{1}{2} \left(\lambda_{min}(\mathbf{Q}) \|\mathbf{e}\|^2 - \frac{4\lambda_{max}^2(\mathbf{P})}{\lambda_{min}(\mathbf{Q})} \varepsilon^2 \right) \\
&\leq -\frac{1}{2} \lambda_{min}(\mathbf{Q}) \|\mathbf{e}\|^2 + \frac{2\lambda_{max}^2(\mathbf{P})}{\lambda_{min}(\mathbf{Q})} \varepsilon^2
\end{aligned}$$

Likewise, part (b) can also be written as

$$\begin{aligned}
\tilde{\mathbf{w}}^T \mathbf{w} - \|\tilde{\mathbf{w}}\|^2 &\leq \|\tilde{\mathbf{w}}\|\|\mathbf{w}\| - \|\tilde{\mathbf{w}}\|^2 \\
&= -\frac{1}{2}(\|\tilde{\mathbf{w}}\| - \|\mathbf{w}\|)^2 - \frac{1}{2}(\|\tilde{\mathbf{w}}\|^2 - \|\mathbf{w}\|^2) \\
&\leq -\frac{1}{2}(\|\tilde{\mathbf{w}}\|^2 - \|\mathbf{w}\|^2)
\end{aligned}$$

Therefore (103) becomes

$$\dot{V} \leq -\frac{1}{2}\lambda_{\min}(\mathbf{Q})\|\mathbf{e}\|^2 + \frac{2\lambda_{\max}^2(\mathbf{P})}{\lambda_{\min}(\mathbf{Q})}\varepsilon^2 - \sigma\|\tilde{\mathbf{w}}\|^2 + \sigma\|\mathbf{w}\|^2$$

$$= \underbrace{-\frac{1}{2}\lambda_{\min}(\mathbf{Q})\|\mathbf{e}\|^2 - \sigma\|\tilde{\mathbf{w}}\|^2}_{(c)} + \sigma\|\mathbf{w}\|^2 + \frac{2\lambda_{\max}^2(\mathbf{P})}{\lambda_{\min}(\mathbf{Q})}\varepsilon^2 \qquad (104)$$

We would like to relate (c) to V by considering

$$V = \mathbf{e}^T\mathbf{P}\mathbf{e} + \tilde{\mathbf{w}}^T\mathbf{\Gamma}\tilde{\mathbf{w}} \leq \lambda_{\max}(\mathbf{P})\|\mathbf{e}\|^2 + \lambda_{\max}(\mathbf{\Gamma})\|\tilde{\mathbf{w}}\|^2 \qquad (105)$$

Now (104) can be further derived as

$$\dot{V} \leq -\alpha V + \left[\alpha\lambda_{\max}(\mathbf{P}) - \frac{\lambda_{\min}(\mathbf{Q})}{2}\right]\|\mathbf{e}\|^2$$

$$+ [\alpha\lambda_{\max}(\mathbf{\Gamma}) - \sigma]\|\tilde{\mathbf{w}}\|^2 + \sigma\|\mathbf{w}\|^2 + \frac{2\lambda_{\max}^2(\mathbf{P})}{\lambda_{\min}(\mathbf{Q})}\varepsilon^2$$

Pick $\alpha \leq \min\left\{\dfrac{\lambda_{\min}(\mathbf{Q})}{2\lambda_{\max}(\mathbf{P})}, \dfrac{\sigma}{\lambda_{\max}(\mathbf{\Gamma})}\right\}$, then we have

$$\dot{V} \leq -\alpha V + \sigma\|\mathbf{w}\|^2 + \frac{2\lambda_{\max}^2(\mathbf{P})}{\lambda_{\min}(\mathbf{Q})}\varepsilon^2 \qquad (106)$$

Hence, $\dot{V} < 0$ whenever

$$(\mathbf{e}, \tilde{\mathbf{w}}) \in E \equiv \left\{ (\mathbf{e}, \tilde{\mathbf{w}}) \middle| V > \frac{1}{\alpha}\left[\sigma\|\mathbf{w}\|^2 + \frac{2\lambda_{\max}^2(\mathbf{P})}{\lambda_{\min}(\mathbf{Q})}\sup_{\tau \geq t_0}\varepsilon^2(\tau)\right]\right\}$$

This implies that $(\mathbf{e}, \tilde{\mathbf{w}})$ is uniformly ultimately bounded. Note that the size of the set E is adjustable by proper selection of α, σ, $\mathbf{P}$, and $\mathbf{Q}$. Smaller size of E implies more accurate in output tracking. However, this parameter adjustment is not always unlimited, because it might induce controller saturation in implementation.

The above derivation only demonstrates the boundedness of the closed loop system, but in practical applications the transient performance is also of great importance. For further development, we may solve the differential inequality in (106) to have the upper bound for the function V

$$V \le e^{-\alpha(t-t_0)}V(t_0) + \frac{\sigma}{\alpha}\|\mathbf{w}\|^2 + \frac{2\lambda_{max}^2(\mathbf{P})}{\alpha\lambda_{min}(\mathbf{Q})}\sup_{t_0 \le \tau \le t}\varepsilon^2(\tau) \qquad (107)$$

By using the definition in (100), we may also find an upper bound of V as

$$V = \mathbf{e}^T\mathbf{P}\mathbf{e} + \tilde{\mathbf{w}}^T\mathbf{\Gamma}\tilde{\mathbf{w}} \ge \lambda_{min}(\mathbf{P})\|\mathbf{e}\|^2 + \lambda_{min}(\mathbf{\Gamma})\|\tilde{\mathbf{w}}\|^2$$

This gives the upper bound for the tracking error

$$\|\mathbf{e}\|^2 \le \frac{1}{\lambda_{min}(\mathbf{P})}[V - \lambda_{min}(\mathbf{\Gamma})\|\tilde{\mathbf{w}}\|^2] \le \frac{1}{\lambda_{min}(\mathbf{P})}V$$

Taking the square root and using (107), we have

$$\|\mathbf{e}\| \le \sqrt{\frac{V(t_0)}{\lambda_{min}(\mathbf{P})}}e^{-\frac{\alpha(t-t_0)}{2}} + \sqrt{\frac{\sigma}{\alpha\lambda_{min}(\mathbf{P})}}\|\mathbf{w}\|$$
$$+ \sqrt{\frac{2\lambda_{max}^2(\mathbf{P})}{\alpha\lambda_{min}(\mathbf{P})\lambda_{min}(\mathbf{Q})}}\sup_{t_0 \le \tau \le t}|\varepsilon(\tau)| \qquad (108)$$

Hence, we have proved that the tracking error is bounded by a weighted exponential function plus a constant. This also implies that by adjusting controller parameters, we may improve output error convergence rate. However, it might also induce controller saturation problem in practice.

The Case When the Bound for the Approximation Error is Known

If the bound for ε is known, i.e., there exists some $\beta > 0$ such that $|\varepsilon| \le \beta$ for all $t \ge t_0$, then u_r in (101) can be modified as

$$u_r = \frac{1+\delta_{max}}{\delta_{min}}|\hat{f}-v|\text{sgn}(\mathbf{b}^T\mathbf{P}\mathbf{e}) + \frac{\beta}{\delta_{min}}\text{sgn}(\mathbf{b}^T\mathbf{P}\mathbf{e})$$

If the control law and the update law are still selected as (97) and (101b) with $\sigma = 0$, then we may have

$$\dot{V} \leq -\mathbf{e}^T \mathbf{Q}\mathbf{e} + 2|\varepsilon|\|\mathbf{b}^T \mathbf{Pe}| - 2\beta|\mathbf{b}^T \mathbf{Pe}|$$
$$\leq -\mathbf{e}^T \mathbf{Q}\mathbf{e} \leq 0$$

Therefore, we may also have asymptotical convergence of the output error by using Barbalat's lemma. It is noted, however, that the assumption for the availability of the bound ε is not practical in most of the applications.

2.9   Backstepping Procedure

Nonlinear control has attracted a lot of attention from researchers and industrial designers in recent years. Many nonlinear controllers outperform their linear counterparts from the viewpoint of the range of operation, rate of convergence and robustness to parameter variations and external disturbances, etc. The stabilization problem of nonlinear systems is much more difficult than that of linear systems. Many techniques have been developed to solve this problem (Isidori, 1989; Slotine and Li, 1991). A recent approach based on passivity designs using a recursive procedure called backstepping was introduced (Kokotovic, 1992; Krstic et al., 1995). The geometric approach is only valid in some local region and with a disturbance-free setting, in general. The backstepping design alleviates some of these limitations.

For system containing uncertainties entering the system in a mismatched fashion, few strategies are available to stabilize the system effectively. The backstepping procedure is perhaps the most well-known design for dealing with mismatched uncertainties. Since in this book we are going to use a backstepping-like method to cover the mismatched uncertainties, a simple introduction of the backstepping procedure is given in this section.

The backstepping design is a systematic procedure which can obtain an explicit global result if the system is in the *strict-feedback form*

$$\dot{x}_i = f_i(x_1, x_2, ..., x_i) + g_i(x_1, x_2, ..., x_i)x_{i+1}, \quad i = 1, ..., n-1$$
$$\dot{x}_n = f_n(x_1, x_2, ..., x_n) + g_n(x_1, x_2, ..., x_n)u \tag{109}$$

The system is called "strict-feedback" because the nonlinear functions f_i and g_i, $i=1,...,n$, are functions of x_j, $j=1,...,i$. For a more general *pure feedback form*,

$$\dot{x}_i = f_i(x_1, x_2, ..., x_{i+1}), \quad i = 1, ..., n-1$$
$$\dot{x}_n = f_n(x_1, x_2, ..., x_n, u) \tag{110}$$

although the results may not be explicit or global, a non-vanishing domain of attraction can still be obtained. The backstepping procedure is not applicable to systems that do not satisfy the pure feedback form. A well-known example is the ball-and-beam system

$$\dot{x}_1 = x_2$$
$$\dot{x}_2 = b(-g \sin x_3 + x_1 x_4^2)$$
$$\dot{x}_3 = x_4$$
$$\dot{x}_4 = u \tag{111}$$

where $b > 0$ is a system parameter. The system is not feedback linearizable, since the relative degree is not well defined. It does not satisfy the pure feedback form either, due to the presence of the term $x_1 x_4^2$ in the second equation above; therefore, the backstepping procedure cannot be applied.

The basic idea of backstepping is to start with a subsystem that can be stabilized with a known feedback law for a known Lyapunov function, and then add an integrator to its input. A new stabilizing feedback law is explicitly designed for the augmented subsystem with a new Lyapunov function. This procedure stops when the control signal comes out.

Let us consider an example introduced in Kokotovic (1992)

$$\dot{x}_1 = f_1(x_1, x_2)$$
$$\dot{x}_2 = x_3 + f_2(x_1, x_2)$$
$$\dot{x}_3 = u + f_3(x_1, x_2, x_3) \tag{112}$$

where f_1, f_2 and f_3 are known smooth functions. It is seen that x_2 is implicitly involved in the first subsystem, while x_3 appears in the second subsystem explicitly. Hence, in the first subsystem, we may see the design concept for pure feedback systems, and in the second subsystem, we can experience the controller construction for strictly feedback systems. The entire backstepping design for (112) is presented in several steps.

STEP 1: Imagine that we are able to stabilize the first subsystem of (112) by regarding x_2 as a control signal. If it is able to be selected as a feedback control $x_2 = v_1(x_1)$ so that

$$\frac{\partial V_1(x_1)}{\partial x_1} f_1(x_1, v_1(x_1)) < 0 \qquad (113)$$

for all $x_1 \neq 0$ and for some known positive definite scalar function $V_1(x_1)$. Actually, x_2 is not a control, and we may not exactly have $x_2 = v_1(x_1)$ for all $t \geq 0$. However, it is possible to define an error signal to measure the difference between x_2 and the feedback law $v_1(x_1)$ as

$$e_2 = x_2 - v_1(e_1) \qquad (114)$$

where $e_1 = x_1$. If it is possible to reduce e_2 asymptotically by proper design of the rest of the part of the control system, then this is a feasible development. Since f_1 in (112) is assumed to be differentiable, the error dynamics of e_1 can be computed as

$$\begin{aligned}
\dot{e}_1 &= f_1(e_1, e_2 + v_1(e_1)) \\
&= f_1(e_1, v_1(e_1)) + e_2 \phi_1(e_1, e_2)
\end{aligned} \qquad (115a)$$

where $\phi_1(e_1, e_2)$ is a known function. On the other hand, the error dynamics of e_2 can be found by taking time derivative of (114) and using (112) to have

$$\dot{e}_2 = x_3 + f_2(e_1, e_2 + v_1(e_1)) - \dot{v}_1(e_1) \qquad (115b)$$

The time derivative of v_1 in (115b) can be calculated with

$$\dot{v}_1 = \frac{\partial v_1}{\partial e_1} f_1(e_1, e_2 + v_1(e_1)) \tag{116}$$

In this step, we obtain the error dynamics for the first and second subsystem represented in (115a) and (115b), respectively. In the next step, we would like to proceed to find the error dynamics of the third subsystem.

STEP 2: Suppose we may select $x_3 = v_2(e_1, e_2)$ so that the system consisting of (115a) and (115b) can be stabilized. Let us define the Lyapunov function candidate

$$V_2(e_1, e_2) = V_1(e_1) + \frac{1}{2} e_2^2 \tag{117}$$

The time derivative of V_2 along the trajectory of (115) can be computed as

$$\dot{V}_2 = \frac{\partial V_1}{\partial e_1} f_1 + e_2 \left[\frac{\partial V_1}{\partial e_1} \phi_1 + e_3 + v_2 + f_2 - \dot{v}_1 \right] \tag{118}$$

where $e_3 = x_3 - v_2(e_1, e_2)$. By selecting

$$v_2 = -e_2 - \frac{\partial V_1}{\partial e_1} \phi_1 - f_2 + \dot{v}_1 \tag{119}$$

equation (118) becomes

$$\dot{V}_2 = \frac{\partial V_1}{\partial e_1} f_1 - e_2^2 + e_2 e_3 \tag{120}$$

According to (113), the first term in (120) is negative, but $\dot{V}_2$ is not negative definite because the term $e_2 e_3$ is indefinite. This term will be cancelled in the next step. Note that we can achieve $x_3 = v_2(e_1, e_2)$ only when the error $e_3 = x_3 - v_2(e_1, e_2)$ converges asymptotically. The error dynamics of e_3 can be found as

$$\dot{e}_3 = u + f_3(e_1, e_2 + v_1, e_3 + v_2) - \dot{v}_2 \tag{121}$$

where $\dot{v}_2$ is known and can be found directly by differentiating (119).

STEP 3: Since u appears in (121) explicitly, we do not need to define a fictitious control any more. Let us construct a Lyapunov function candidate as

$$V_3(e_1, e_2, e_3) = V_2(e_1, e_2) + \frac{1}{2} e_3^2 \tag{122}$$

and its time derivative along the trajectory of the error dynamics consisting of (115) and (121) is

$$\dot{V}_3 = \frac{\partial V_1}{\partial e_1} f_1 - e_2^2 + e_3(u + e_2 + f_3 - \dot{v}_2) \tag{123}$$

Therefore, if we choose

$$u = -e_3 - e_2 - f_3 + \dot{v}_2 \tag{124}$$

then (123) becomes

$$\dot{V}_3 = \frac{\partial V_1}{\partial e_1} f_1 - e_2^2 - e_3^2 < 0 \tag{125}$$

Hence, the closed loop system is asymptotic stable.

In the above derivation, the system is assumed to be known. The adaptive backstepping and robust backstepping have all been well-developed (Krstic et al., 1995) for systems containing uncertainties.

2.10 Multiple-Surface Sliding Control

The sliding control is effective in controlling nonlinear system containing matched uncertainties. When some uncertainties enter the system in a mismatched fashion, a backstepping-like design called the multiple-surface sliding control (Won and Hedrick 1996; Yip and Hedrick 1997; Gerdes and Hedrick 1999) can be applied.

Let us consider a control system described by the equations

$$
\begin{aligned}
\dot{x}_1 &= x_2 + f_1(\mathbf{x},t) \\
\dot{x}_2 &= x_3 + f_2(\mathbf{x},t) \\
&\vdots \\
\dot{x}_n &= g(\mathbf{x},t)u + f_n(\mathbf{x},t) \\
y &= x_1
\end{aligned}
\tag{126}
$$

where $\mathbf{x} = [x_1 \cdots x_n]^T \in \Re^n$ is a vector of measurable states, $u \in \Re$ and $y \in \Re$ are input and output signals respectively. The functions $f_i(\mathbf{x},t) \in \Re$, $i = 1, \cdots, n$ and $g(\mathbf{x},t) \in \Re$ are unknown functions but bounds of their variations should be available. The control gain function $g(\mathbf{x},t)$ is assumed to be non-singular for all admissible $\mathbf{x}$ and for all time t. Let us assume that $f_i(\mathbf{x},t)$ and $g(\mathbf{x},t)$ are modeled as

$$
\begin{aligned}
f_i &= f_{mi} + \Delta f_i, \quad i = 1, \ldots, n \\
g &= g_m \Delta g
\end{aligned}
$$

where f_{mi} and g_m are known nominal values of f_i and g, respectively. The uncertain terms Δf_i and Δg are assumed to be bounded as

$$
|\Delta f_i| \le \alpha_i, \quad i = 1, \ldots, n
$$

$$
0 \le \gamma_{\min} = \frac{g_{\min}}{g_m} \le \Delta g \le \frac{g_{\max}}{g_m} = \gamma_{\max}
$$

for some known constants $\alpha_i > 0$, $g_{\min} > 0$ and $g_{\max} > 0$. We would like to design a tracking controller so that the output y tracks the desired trajectory y_d regardless of the presence of uncertainties. Since most of uncertainties in (126) do not satisfy the *matching condition*, traditional sliding control law is not applicable. The multiple surface sliding controller is designed in several steps:

STEP 1: Define n sliding surfaces as $s_i = x_i - x_{id}$, $i = 1, \ldots, n$, where x_{id} represents the desired trajectory of state x_i. Let $x_{1d} = y_d$, and calculate the time derivative of s_1 as

$$\begin{aligned} s_1 &= \dot{x}_1 - \dot{x}_{1d} \\ &= x_2 + f_1 - \dot{y}_d \\ &= (s_2 + x_{2d}) + (f_{m1} + \Delta f_1) - \dot{y}_d \end{aligned} \tag{127}$$

The signal x_{2d} can be regarded as a virtual control in the dynamic system represented in (127), and we can pick it as

$$x_{2d} = -f_{m1} + \dot{y}_d - c_1(x_1)\frac{s_1}{\phi_1} \tag{128}$$

where $c_1(x_1) > 0$ is a design parameter to be determined later and $\phi_1 > 0$ is the thickness of the boundary layer of the sliding surface $s_1 = 0$. With these selections, equation (127) becomes

$$\dot{s}_1 = s_2 + \Delta f_1 - c_1\frac{s_1}{\phi_1} \tag{129}$$

STEP 2: Take the time derivative of the error signal s_2 to have

$$\dot{s}_2 = (s_3 + x_{3d}) + f_2 - \dot{x}_{2d} \tag{130}$$

The term $\dot{x}_{2d}$ can be obtained from differentiation of (128) as

$$\begin{aligned} \dot{x}_{2d} &= -\dot{f}_{m1} + \ddot{y}_d - \frac{\dot{c}_1}{\phi_1}s_1 - \frac{c_1}{\phi_1}\dot{s}_1 \\ &= -\dot{f}_{m1} + \ddot{y}_d - \frac{dc_1}{dx_1}x_2\frac{s_1}{\phi_1} - \frac{c_1}{\phi_1}\left(s_2 - c_1\frac{s_1}{\phi_1}\right) \\ &\quad - \frac{dc_1}{dx_1}\frac{s_1}{\phi_1}f_{m1} - \left(\frac{dc_1}{dx_1}\frac{s_1}{\phi_1} + \frac{c_1}{\phi_1}\right)\Delta f_1 \end{aligned}$$

Define $\dot{x}_{2dk} = -\dot{f}_{m1} + \ddot{y}_d - \dfrac{dc_1}{dx_1} x_2 \dfrac{s_1}{\phi_1} - \dfrac{c_1}{\phi_1}\left(s_2 - c_1\dfrac{s_1}{\phi_1}\right) - \dfrac{dc_1}{dx_1}\dfrac{s_1}{\phi_1}f_{m1}$ as the

known part of $\dot{x}_{2d}$ and $\dot{x}_{2du} = -\left(\dfrac{dc_1}{dx_1}\dfrac{s_1}{\phi_1} + \dfrac{c_1}{\phi_1}\right)\Delta f$ is the unknown part, and

the above equation can be expressed as

$$\dot{x}_{2d} = \dot{x}_{2dk} + \dot{x}_{2du} \tag{131}$$

Note that $\dot{x}_{2du}$ is not known but its variation bound is assumed to be

available as $\beta_2 = \left|\dfrac{dc_1}{dx_1}\dfrac{s_1}{\phi_1} + \dfrac{c_1}{\phi_1}\right|\alpha_1 > 0$ and we may have $|\dot{x}_{2du}| \le \beta_2$. Let us

regard x_{3d} in (130) as a virtual control, and we can have the design

$$x_{3d} = -f_{m2} + \dot{x}_{2dk} - c_2(x_1, x_2)\dfrac{s_2}{\phi_2} \tag{132}$$

where $c_2(x_1, x_2) > 0$ is a design parameter to be determined later, and
$\phi_2 > 0$ is the thickness of sliding surface $s_2 = 0$. With this selection,
equation (130) becomes

$$\dot{s}_2 = s_3 + \Delta f_2 - \dot{x}_{2du} - c_2\dfrac{s_2}{\phi_2} \tag{133}$$

STEP i: $i = 2,..., n-1$

Following the same procedure as derived above, we may have
the results for the i-th subsystem $i = 2,..., n-1$ as

$$\dot{s}_i = s_{i+1} + \Delta f_i - \dot{x}_{idu} - c_i(x_1,..., x_i)\dfrac{s_i}{\phi_i} \tag{134}$$

$$x_{(i+1)d} = -f_{mi} + \dot{x}_{idk} - c_i(x_1,..., x_i)\dfrac{s_i}{\phi_i} \tag{135}$$

$$\dot{x}_{(i+1)d} = \dot{x}_{(i+1)dk} + \dot{x}_{(i+1)du} \tag{136}$$

$$|\dot{x}_{idu}| \le \beta_i \tag{137}$$

STEP n:

By taking the time derivative of s_n and using the last line of (126), we have the error dynamics for the last subsystem as

$$\dot{s}_n = f_n + gu - \dot{x}_{ndk} - \dot{x}_{ndu} \tag{138}$$

It is seen that the control u appears explicitly which is going to be designed to stabilize the system. Now, let us collect all the error dynamics together from (129), (133), (134) and (138) to form

$$\dot{s}_1 = s_2 + \Delta f_1 - c_1 \frac{s_1}{\phi_1}$$

$$\dot{s}_2 = s_3 + \Delta f_2 - \dot{x}_{2du} - c_2 \frac{s_2}{\phi_2}$$

$$\vdots$$

$$\dot{s}_{n-1} = s_n + \Delta f_{n-1} - \dot{x}_{(n-1)du} - c_{n-1}(x_1,...,x_{n-1}) \frac{s_{n-1}}{\phi_{n-1}}$$

$$\dot{s}_n = f_n + gu - \dot{x}_{ndk} - \dot{x}_{ndu}$$

If we may find a controller u in (138) so that s_n converges, then s_{n-1} in (134) will converge if $c_{n-1}(x_1,...,x_{n-1})$ is properly selected to cover the bounded uncertainties $\Delta f_{n-1} - \dot{x}_{(n-1)du}$. Likewise, for small s_3 in (133), if we may design $c_2(x_1,x_2)$ so that bounded uncertainties $\Delta f_2 - \dot{x}_{2du}$ is properly covered, then we may have convergence of s_2. Finally, if $c_1(x_1)$ in (129) is selected to cover the uncertainty Δf_1, then we may have the convergence of s_1. To design u and $c_i, i = 1,...,n-1$, we need to apply the Lyapunov direct method starting from the n-th surface to the first surface. So, let us define a Lyapunov function candidate for the n-th surface as

$$V_n = \frac{1}{2} s_n^2 \tag{139}$$

Its time derivative along the trajectory of (138) can be calculated as

$$\dot{V}_n = s_n (f_n + gu - \dot{x}_{ndk} - \dot{x}_{ndu}) \tag{140}$$

The control signal is selected to be

$$u = \frac{1}{g_m} \left[-f_{mn} + \dot{x}_{ndk} - c_n(\mathbf{x}) \frac{s_n}{\phi_n} \right] \tag{141}$$

where $c_n(\mathbf{x})$ is to be determined. Substituting (141) into (140), we have

$$\dot{V}_n = s_n \left[(1 - \Delta g)(f_{mn} - \dot{x}_{ndk}) + (\Delta f_n - \dot{x}_{ndu}) - \Delta g c_n \frac{s_n}{\phi_n} \right]$$

$$\leq (1 - \gamma_{\min}) \left| f_{mn} - \dot{x}_{ndk} \right| \left| s_n \right| + (\alpha_n + \beta_n) \left| s_n \right| - \gamma_{\min} c_n \frac{s_n^2}{\phi_n} \tag{142}$$

Choose $c_n = \dfrac{1}{\gamma_{\min}} [(1 - \gamma_{\min}) \left| f_{mn} - \dot{x}_{ndk} \right| + (\alpha_n + \beta_n) + \eta_n]$, where $\eta_n > 0$.

Then following the same reasoning in (32), for $\left| s_n \right| > \phi_n$, inequality (142) becomes

$$\dot{V}_n \leq -\eta_n \frac{s_n^2}{\phi_n} \tag{143}$$

Hence, the boundary layer of $s_n = 0$ is attractive, and we have $\left| s_n \right| \leq \phi_n$ in finite time. Timing for this convergence will be detailed in chapter 4 where adaptive version of the multiple-surface sliding controller is derived. Next, let us define

$$V_{n-1} = \frac{1}{2} s_{n-1}^2 \tag{144}$$

The time derivative of V_{n-1} can be found as

$$\dot{V}_{n-1} = s_{n-1} \left[s_n + \Delta f_{n-1} - \dot{x}_{(n-1)du} - c_{n-1} \frac{s_{n-1}}{\phi_{n-1}} \right]$$

$$\leq |s_{n-1}| (\phi_n + \alpha_{n-1} + \beta_{n-1}) - c_{n-1} \frac{s_{n-1}^2}{\phi_{n-1}}$$

(145)

By picking $c_{n-1} = \phi_n + \alpha_{n-1} + \beta_{n-1} + \eta_{n-1}$, with $\eta_{n-1} > 0$, inequality (145) becomes

$$\dot{V}_{n-1} \leq -\eta_{n-1} \frac{s_{n-1}^2}{\phi_{n-1}}$$

(146)

Let $V_i = \frac{1}{2} s_i^2$ for all $i = 1, \ldots, n-1$, then we have

$$\dot{V}_i \leq |s_i| (\phi_{i+1} + \alpha_i + \beta_i) - c_i \frac{s_i^2}{\phi_i}$$

(147)

We may pick $c_i = \phi_{i+1} + \alpha_i + \beta_i + \eta_i$ to have

$$\dot{V}_i \leq -\eta_i \frac{s_i^2}{\phi_i}$$

(148)

and $|s_i| \leq \phi_i$ as $t \to \infty$.

Chapter 3

Underactuated System Dynamics and Coordinate Transformation

A coordinate transformation is introduced in this chapter to decouple the control input so that an underactuated system can be represented into a special cascade form. The dynamic equations for mechanical systems are reviewed in section 3.1, and the underactuated system is formally defined in section 3.2. The collocated partial feedback linearization is used in section 3.3 to linearize the actuated part of the system into a double integrator such that the coordinate transformation can be applied. We firstly introduce the coordinate transformation for a general underactuated system in section 3.4. Since a set of partial differential equations are to be solved for obtaining the diffeomorphism, to find the solution is usually not easy even though they are proved to be completely integrable. Explicit transformations for 2-DOF systems are given in section 3.5 to facilitate the formation of the special cascade form. The decoupling of high-order systems is introduced in section 3.6.

3.1 Dynamics of Mechanical Systems

In this section, we review the derivation of equations of motion for mechanical systems. Let Q be an n-dimensional configuration manifold, and $\mathbf{q} \in Q$ is a vector of generalized coordinates. For a Lagrangian system in the Euclidean space, we have $Q = \Re^n$. The *kinetic energy* is defined as a quadratic function of the generalized velocity vector $\dot{\mathbf{q}} = [\dot{q}_1 \quad \cdots \quad \dot{q}_n]^T \in \Re^n$ in the form

$$K = \frac{1}{2} \sum_{i,j}^{n} m_{ij}(\mathbf{q}) \dot{q}_i \dot{q}_j = \frac{1}{2} \dot{\mathbf{q}}^T \mathbf{M}(\mathbf{q}) \dot{\mathbf{q}} \tag{1}$$

where $\mathbf{M}(\mathbf{q}) = \mathbf{M}^T(\mathbf{q}) \in \Re^{n \times n}$ is a positive definite inertia matrix, and $m_{ij} \in \Re^+$, $i = 1,...,n$ and $j = 1,...,n$, are entries of $\mathbf{M}$. The *potential energy* of the mechanical system in the configuration manifold is independent to the generalized velocity vector and can be denoted as $V(\mathbf{q}) \in \Re$. Then the *Lagrangian* is defined to be

$$L = K - V = \frac{1}{2} \sum_{i,j}^{n} m_{ij}(\mathbf{q}) \dot{q}_i \dot{q}_j - V(\mathbf{q}) \tag{2}$$

The Euler-Lagrange equations of motion of a mechanical system under external control force is well-known to be given by

$$\frac{d}{dt} \frac{\partial L}{\partial \dot{\mathbf{q}}} - \frac{\partial L}{\partial \mathbf{q}} = \mathbf{B}(\mathbf{q})\mathbf{u} \tag{3}$$

where $\mathbf{B}(\mathbf{q}) \in \Re^{n \times m}$ is a coefficient matrix for the control force vector $\mathbf{u} \in \Re^m$. Let us look at (3) term by term. To take the partial derivative $\dfrac{\partial L}{\partial \dot{\mathbf{q}}}$ in the first term of (3), let us consider the partial differentiation of the Lagrangian in (2) with respect to the k-th entry of the generalized velocity vector first

$$\frac{\partial L}{\partial \dot{q}_k} = \sum_j m_{kj}(\mathbf{q}) \dot{q}_j \tag{4}$$

Thus we may take the time derivative of (4) to have

$$\frac{d}{dt}\frac{\partial L}{\partial \dot{q}_k} = \sum_j \frac{dm_{kj}(\mathbf{q})}{dt}\dot{q}_j + \sum_j m_{kj}(\mathbf{q})\ddot{q}_j$$

$$= \sum_{i,j}\frac{\partial m_{kj}}{\partial q_i}\dot{q}_i\dot{q}_j + \sum_j m_{kj}(\mathbf{q})\ddot{q}_j \qquad (5)$$

$$= \frac{1}{2}\sum_{i,j}\left[\frac{\partial m_{kj}}{\partial q_i} + \frac{\partial m_{ki}}{\partial q_j}\right]\dot{q}_i\dot{q}_j + \sum_j m_{kj}(\mathbf{q})\ddot{q}_j$$

where the last line is obtained by interchanging the order of the summation and using the symmetry property. To compute the second term $\frac{\partial L}{\partial \mathbf{q}}$ in (3), we also consider the partial differentiation of the Lagrangian with respect to the k-th entry of the generalized coordinate vector first

$$\frac{\partial L}{\partial q_k} = \frac{\partial}{\partial q_k}\left[\frac{1}{2}\sum_{i,j}^{n} m_{ij}(\mathbf{q})\dot{q}_i\dot{q}_j - V(\mathbf{q})\right]$$

$$= \frac{1}{2}\sum_{i,j}^{n}\frac{\partial m_{ij}}{\partial q_k}\dot{q}_i\dot{q}_j - \frac{\partial V}{\partial q_k} \qquad (6)$$

By using (5) and (6), the component form of the forced Euler-Lagrange equation in (3) can be represented as

$$\sum_j m_{kj}(\mathbf{q})\ddot{q}_j + \frac{1}{2}\sum_{i,j}\left[\frac{\partial m_{kj}}{\partial q_i} + \frac{\partial m_{ki}}{\partial q_j} - \frac{\partial m_{ij}}{\partial q_k}\right]\dot{q}_i\dot{q}_j + \frac{\partial V}{\partial q_k} = \mathbf{e}_k^T\mathbf{B}(\mathbf{q})\mathbf{u} \qquad (7)$$

Where $k = 1,...,n$ and $\mathbf{e}_k \in \Re^n$ is the k-th basis of $\Re^n$. This equation of motion can be further written to

$$\sum_j m_{kj}(\mathbf{q})\ddot{q}_j + \sum_{i,j} c_{ijk}(\mathbf{q})\dot{q}_i\dot{q}_j + \frac{\partial V}{\partial q_k} = \mathbf{e}_k^T\mathbf{B}(\mathbf{q})\mathbf{u} \qquad (8)$$

where $c_{ijk}(\mathbf{q}) = \dfrac{1}{2}\left[\dfrac{\partial m_{kj}}{\partial q_i} + \dfrac{\partial m_{ki}}{\partial q_j} - \dfrac{\partial m_{ij}}{\partial q_k} \right]$ is well-known to be the

Christoffel symbols of the first kind. Equation (8) can thus be rewritten in the vector form

$$\mathbf{M}(\mathbf{q})\ddot{\mathbf{q}} + \mathbf{C}(\mathbf{q},\dot{\mathbf{q}})\dot{\mathbf{q}} + \mathbf{g}(\mathbf{q}) = \mathbf{B}(\mathbf{q})\mathbf{u} \qquad (9)$$

where $\mathbf{C}(\mathbf{q},\dot{\mathbf{q}})\dot{\mathbf{q}} \in \mathfrak{R}^n$ contains the centrifugal and Coriolis effects and $\mathbf{g}(\mathbf{q}) \in \mathfrak{R}^n$ is the gravitation vector.

3.2 Underactuated Systems

Dimension of the matrix $\mathbf{B}(\mathbf{q}) \in \mathfrak{R}^{n \times m}$ in (9) relates to the property of the system. If $m=n$ and $\mathbf{B}(\mathbf{q})$ is invertible, then the system is *fully-actuated*. In other words, we may roughly say that a mechanical system is fully-actuated if the number of control inputs is equal to the dimension of its configuration manifold. Actually, we still need the invertibility of $\mathbf{B}(\mathbf{q})$ for all $\mathbf{q} \in Q$ to ensure the system to be fully actuated. Suppose $\mathbf{B}(\mathbf{q})$ is invertible and under the assumption that $\mathbf{M}(\mathbf{q})$, $\mathbf{C}(\mathbf{q},\dot{\mathbf{q}})$ and $\mathbf{g}(\mathbf{q})$ are available, we may select the control law

$$\mathbf{u} = \mathbf{B}^{-1}(\mathbf{q})[\mathbf{C}(\mathbf{q},\dot{\mathbf{q}})\dot{\mathbf{q}} + \mathbf{g}(\mathbf{q}) + \mathbf{M}(\mathbf{q})\mathbf{v}] \qquad (10)$$

where $\mathbf{v} \in \mathfrak{R}^n$ is some control signal to be designed so that the closed loop system becomes a linear time-invariant dynamics

$$\ddot{\mathbf{q}} = \mathbf{v} \qquad (11)$$

This proves that fully-actuated mechanical systems are feedback linearizable without internal dynamics. The control law (10) is feasible if all parameters in (9) are available. However, in practical applications, the system model inevitably contains various uncertainties and external disturbances; therefore, the control in (10) cannot be applied directly.

On the other hand, if $m < n$, then (9) is *underactuated*. Since m can be regarded as the number of actuators while n is the number of the configuration variables, so we may say that an underactuated mechanical system is a system with fewer actuators than its configuration variables. Because $m \neq n$, the matrix $\mathbf{B(q)}$ is not possible to be invertible, and the control law in (10) cannot be used. This implies that underactuated mechanical systems are not exact feedback linearizable. Since the actuators are fewer, the controllability of the system is not always satisfied. In this book, we assume that all systems considered are controllable.

Suppose (9) can be arranged so that $\mathbf{B(q)}$ is in the form

$$\mathbf{B(q)} = \begin{bmatrix} \mathbf{0}_{(n-m) \times m} \\ \mathbf{I}_m \end{bmatrix} \tag{12}$$

We call the first n-m equations in (9) to be the *unactuated subsystem* while the last m equations *actuated subsystem*. We may thus represent $\mathbf{q}$ to be

$$\mathbf{q} = \begin{bmatrix} \mathbf{q}_u \\ \mathbf{q}_a \end{bmatrix} \begin{matrix} \}(n-m) \times 1 \\ \}m \times 1 \end{matrix} \tag{13}$$

where $\mathbf{q}_u \in \mathfrak{R}^{n-m}$ is the vector of unactuated configuration variables, and $\mathbf{q}_a \in \mathfrak{R}^m$ is the vector of actuated configuration variables. Thus (9) can be rewritten as

$$\begin{bmatrix} \mathbf{m}_{11}(\mathbf{q}) & \mathbf{m}_{12}(\mathbf{q}) \\ \mathbf{m}_{21}(\mathbf{q}) & \mathbf{m}_{22}(\mathbf{q}) \end{bmatrix} \begin{bmatrix} \ddot{\mathbf{q}}_u \\ \ddot{\mathbf{q}}_a \end{bmatrix} + \begin{bmatrix} \mathbf{c}_{11}(\mathbf{q},\dot{\mathbf{q}}) & \mathbf{c}_{12}(\mathbf{q},\dot{\mathbf{q}}) \\ \mathbf{c}_{21}(\mathbf{q},\dot{\mathbf{q}}) & \mathbf{c}_{22}(\mathbf{q},\dot{\mathbf{q}}) \end{bmatrix} \begin{bmatrix} \dot{\mathbf{q}}_u \\ \dot{\mathbf{q}}_a \end{bmatrix} + \begin{bmatrix} \mathbf{g}_1(\mathbf{q}) \\ \mathbf{g}_2(\mathbf{q}) \end{bmatrix} = \begin{bmatrix} \mathbf{0} \\ \mathbf{u}_2 \end{bmatrix} \tag{14}$$

We may also represent (14) in the component form

$$\mathbf{m}_{11}(\mathbf{q})\ddot{\mathbf{q}}_u + \mathbf{m}_{12}(\mathbf{q})\ddot{\mathbf{q}}_a + \mathbf{c}_{11}(\mathbf{q},\dot{\mathbf{q}})\dot{\mathbf{q}}_u + \mathbf{c}_{12}(\mathbf{q},\dot{\mathbf{q}})\dot{\mathbf{q}}_a + \mathbf{g}_1(\mathbf{q}) = \mathbf{0} \tag{15a}$$

$$\mathbf{m}_{21}(\mathbf{q})\ddot{\mathbf{q}}_u + \mathbf{m}_{22}(\mathbf{q})\ddot{\mathbf{q}}_a + \mathbf{c}_{21}(\mathbf{q},\dot{\mathbf{q}})\dot{\mathbf{q}}_u + \mathbf{c}_{22}(\mathbf{q},\dot{\mathbf{q}})\dot{\mathbf{q}}_a + \mathbf{g}_2(\mathbf{q}) = \mathbf{u}_2 \tag{15b}$$

It is seen that there is no control in (15a) and hence we may not fully linearize this system. However, it is still possible to partially linearize (15b) into a double integrator. One of these linearization methods is the well-known collocated partial feedback linearization proposed by Spong (1996).

3.3 Collocated Partial Feedback Linearization

In the following derivation, we use a simplified version of (15) as

$$\mathbf{m}_{11}(\mathbf{q})\ddot{\mathbf{q}}_u + \mathbf{m}_{12}(\mathbf{q})\ddot{\mathbf{q}}_a + \mathbf{h}_1(\mathbf{q},\dot{\mathbf{q}}) = 0 \qquad (16a)$$

$$\mathbf{m}_{21}(\mathbf{q})\ddot{\mathbf{q}}_u + \mathbf{m}_{22}(\mathbf{q})\ddot{\mathbf{q}}_a + \mathbf{h}_2(\mathbf{q},\dot{\mathbf{q}}) = \mathbf{u}_2 \qquad (16b)$$

where nonlinear functions $\mathbf{h}_1(\mathbf{q},\dot{\mathbf{q}})$ and $\mathbf{h}_2(\mathbf{q},\dot{\mathbf{q}})$ are defined as

$$\mathbf{h}_1(\mathbf{q},\dot{\mathbf{q}}) = \mathbf{c}_{11}(\mathbf{q},\dot{\mathbf{q}})\dot{\mathbf{q}}_u + \mathbf{c}_{12}(\mathbf{q},\dot{\mathbf{q}})\dot{\mathbf{q}}_a + \mathbf{g}_1(\mathbf{q})$$

$$\mathbf{h}_2(\mathbf{q},\dot{\mathbf{q}}) = \mathbf{c}_{21}(\mathbf{q},\dot{\mathbf{q}})\dot{\mathbf{q}}_u + \mathbf{c}_{22}(\mathbf{q},\dot{\mathbf{q}})\dot{\mathbf{q}}_a + \mathbf{g}_2(\mathbf{q})$$

From (16a), we may compute $\ddot{\mathbf{q}}_u$ to be

$$\ddot{\mathbf{q}}_u = -\mathbf{m}_{11}^{-1}\mathbf{m}_{12}\ddot{\mathbf{q}}_a - \mathbf{m}_{11}^{-1}\mathbf{h}_1 \qquad (17)$$

where $\mathbf{m}_{11}$ is assumed to be invertible. Substitute (17) into (16b) to give

$$\mathbf{m}_{21}(-\mathbf{m}_{11}^{-1}\mathbf{m}_{12}\ddot{\mathbf{q}}_a - \mathbf{m}_{11}^{-1}\mathbf{h}_1) + \mathbf{m}_{22}\ddot{\mathbf{q}}_a + \mathbf{h}_2 = \mathbf{u}_2$$

This further implies

$$(\mathbf{m}_{22} - \mathbf{m}_{21}\mathbf{m}_{11}^{-1}\mathbf{m}_{12})\ddot{\mathbf{q}}_a - \mathbf{m}_{21}\mathbf{m}_{11}^{-1}\mathbf{h}_1 + \mathbf{h}_2 = \mathbf{u}_2$$

Define $\alpha(\mathbf{q}) = \mathbf{m}_{22} - \mathbf{m}_{21}\mathbf{m}_{11}^{-1}\mathbf{m}_{12}$ and assume that it is non-singular, then we have

$$\ddot{\mathbf{q}}_a + \underbrace{(\mathbf{m}_{22} - \mathbf{m}_{21}\mathbf{m}_{11}^{-1}\mathbf{m}_{12})^{-1}(\mathbf{h}_2 - \mathbf{m}_{21}\mathbf{m}_{11}^{-1}\mathbf{h}_1)}_{\beta(\mathbf{q},\dot{\mathbf{q}})} = \underbrace{(\mathbf{m}_{22} - \mathbf{m}_{21}\mathbf{m}_{11}^{-1}\mathbf{m}_{12})^{-1}}_{\alpha(\mathbf{q})}\mathbf{u}_2$$

Define $\beta(\mathbf{q},\dot{\mathbf{q}}) = (\mathbf{m}_{22} - \mathbf{m}_{21}\mathbf{m}_{11}^{-1}\mathbf{m}_{12})^{-1}(\mathbf{h}_2 - \mathbf{m}_{21}\mathbf{m}_{11}^{-1}\mathbf{h}_1)$ to simplify the equation to

$$\ddot{\mathbf{q}}_a + \beta(\mathbf{q},\dot{\mathbf{q}}) = \alpha^{-1}(\mathbf{q})\mathbf{u}_2 \tag{18}$$

Since all system parameters are known, we may thus select

$$\mathbf{u}_2 = \alpha(\mathbf{q})[\mathbf{u} + \beta(\mathbf{q},\dot{\mathbf{q}})] \tag{19}$$

so that (18) becomes a double integrator

$$\ddot{\mathbf{q}}_a = \mathbf{u} \tag{20}$$

This is a linearization of the actuated part of the system. Put (20) into (17), we have

$$\ddot{\mathbf{q}}_u = \underbrace{- \mathbf{m}_{11}^{-1}\mathbf{m}_{12}}_{G_0(\mathbf{q})}\mathbf{u} \underbrace{- \mathbf{m}_{11}^{-1}\mathbf{h}_1}_{-f_0(\mathbf{q},\dot{\mathbf{q}})}$$

By defining $\mathbf{f}_0(\mathbf{q},\dot{\mathbf{q}}) = -\mathbf{m}_{11}^{-1}\mathbf{h}_1$ and $\mathbf{G}_0(\mathbf{q}) = -\mathbf{m}_{11}^{-1}\mathbf{m}_{12}$ to give

$$\ddot{\mathbf{q}}_u = \mathbf{f}_0(\mathbf{q},\dot{\mathbf{q}}) + \mathbf{G}_0(\mathbf{q})\mathbf{u} \tag{21}$$

From (21) and (20), we have the partially linearized system equation

$$\begin{aligned}\ddot{\mathbf{q}}_u &= \mathbf{f}_0(\mathbf{q},\dot{\mathbf{q}}) + \mathbf{G}_0(\mathbf{q})\mathbf{u} \\ \ddot{\mathbf{q}}_a &= \mathbf{u}\end{aligned} \tag{22}$$

It is seen that the actuated part of the system is linearized by the control (19) into a double integrator while the unactuated part is still complex. The underactuated nature of the system can be observed from the presence of $\mathbf{u}$ in both subsystems in (22). This largely increases the difficulty in the controller design.

There are several other representations for (22). For example, we may rewrite (22) into the vector form as

$$\begin{bmatrix}\ddot{\mathbf{q}}_u \\ \ddot{\mathbf{q}}_a\end{bmatrix} = \begin{bmatrix}\mathbf{f}_0(\mathbf{q},\dot{\mathbf{q}}) \\ \mathbf{0}\end{bmatrix} + \begin{bmatrix}\mathbf{G}_0(\mathbf{q}) \\ \mathbf{I}_m\end{bmatrix}\mathbf{u} \tag{23}$$

The other representation can be obtained by defining the state vector

$$\mathbf{x} = [\mathbf{x}_1^T \quad \mathbf{x}_2^T \quad \mathbf{x}_3^T \quad \mathbf{x}_4^T]^T = [\mathbf{q}_u^T \quad \dot{\mathbf{q}}_u^T \quad \mathbf{q}_a^T \quad \dot{\mathbf{q}}_a^T]^T \in \mathfrak{R}^{2n}$$

and we may have the state space representation of (22) as

$$\begin{aligned}
\dot{\mathbf{x}}_1 &= \mathbf{x}_2 \\
\dot{\mathbf{x}}_2 &= \mathbf{f}_0(\mathbf{x}) + \mathbf{G}_0(\mathbf{x})\mathbf{u} \\
\dot{\mathbf{x}}_3 &= \mathbf{x}_4 \\
\dot{\mathbf{x}}_4 &= \mathbf{u}
\end{aligned} \tag{24}$$

This can also be written into the vector form as

$$\begin{bmatrix} \dot{\mathbf{x}}_1 \\ \dot{\mathbf{x}}_2 \\ \dot{\mathbf{x}}_3 \\ \dot{\mathbf{x}}_4 \end{bmatrix} = \begin{bmatrix} \mathbf{x}_2 \\ \mathbf{f}_0(\mathbf{x}) \\ \mathbf{x}_4 \\ \mathbf{0} \end{bmatrix} + \begin{bmatrix} \mathbf{0} \\ \mathbf{G}_0(\mathbf{x}) \\ \mathbf{0} \\ \mathbf{I}_m \end{bmatrix} \mathbf{u} \tag{25}$$

3.4 Decoupling of General Underactuated Systems

In this section, we are going to go through a formal procedure to decouple an underactuated system in (24) into the form

$$\begin{aligned}
\dot{\mathbf{z}}_1 &= \mathbf{f}_1(\mathbf{z}) \\
\dot{\mathbf{z}}_2 &= \mathbf{f}_2(\mathbf{z}) \\
\dot{\mathbf{z}}_3 &= \mathbf{z}_4 \\
\dot{\mathbf{z}}_4 &= \mathbf{u}
\end{aligned} \tag{26}$$

where $\mathbf{z}_1, \mathbf{z}_2 \in \mathfrak{R}^{n-m}$ and $\mathbf{z}_3, \mathbf{z}_4 \in \mathfrak{R}^m$. It is seen that the last two lines in both (24) and (26) are the same, so we don't really need to find their transformations. However, in the first two equations in (26), there is no presence of $\mathbf{u}$ and we may say that the underactuated dynamics in (24) has been decoupled into a special cascade equation in (26). To find the transformation, let $\mathbf{\Phi}: \mathfrak{R}^{2n} \rightarrow \mathfrak{R}^{2n}$ be a diffeomorphism which maps (24) into (26) with the coordinate transformation

$$\mathbf{z} = \mathbf{\Phi}(\mathbf{x}) \tag{27}$$

The component form of $\boldsymbol{\Phi}(\mathbf{x})$ can be represented as

$$\boldsymbol{\Phi}(\mathbf{x}) = \begin{bmatrix} \boldsymbol{\Phi}_1(\mathbf{x}) \\ \boldsymbol{\Phi}_2(\mathbf{x}) \\ \boldsymbol{\Phi}_3(\mathbf{x}) \\ \boldsymbol{\Phi}_4(\mathbf{x}) \end{bmatrix}$$

where $\boldsymbol{\Phi}_1, \boldsymbol{\Phi}_2 \in \Re^{n-m}$ and $\boldsymbol{\Phi}_3, \boldsymbol{\Phi}_4 \in \Re^m$ are defined as

$$\boldsymbol{\Phi}_1(\mathbf{x}) = \begin{bmatrix} \phi_1(\mathbf{x}) \\ \vdots \\ \phi_{n-m}(\mathbf{x}) \end{bmatrix} \quad \boldsymbol{\Phi}_2(\mathbf{x}) = \begin{bmatrix} \phi_{n-m+1}(\mathbf{x}) \\ \vdots \\ \phi_{2(n-m)}(\mathbf{x}) \end{bmatrix}$$

$$\boldsymbol{\Phi}_3(\mathbf{x}) = \begin{bmatrix} \phi_{2(n-m)+1}(\mathbf{x}) \\ \vdots \\ \phi_{2(n-m)+m}(\mathbf{x}) \end{bmatrix} \quad \boldsymbol{\Phi}_4(\mathbf{x}) = \begin{bmatrix} \phi_{2(n-m)+m+1}(\mathbf{x}) \\ \vdots \\ \phi_{2n}(\mathbf{x}) \end{bmatrix}$$

with $\phi_i : \Re^n \rightarrow \Re$, $i = 1,...,2n$. Now, let us take the time derivative to the both sides of (27) as

$$\dot{\mathbf{z}} = \frac{\partial \boldsymbol{\Phi}(\mathbf{x})}{\partial \mathbf{x}} \dot{\mathbf{x}} \tag{28}$$

Using (24) and (26), we may rewrite (28) into

$$\begin{bmatrix} \mathbf{f}_1(\mathbf{z}) \\ \mathbf{f}_2(\mathbf{z}) \\ \mathbf{z}_4 \\ \mathbf{u} \end{bmatrix}_{2n \times 1} = \begin{bmatrix} \dfrac{\partial \boldsymbol{\Phi}_1}{\partial \mathbf{x}_1} & \dfrac{\partial \boldsymbol{\Phi}_1}{\partial \mathbf{x}_2} & \dfrac{\partial \boldsymbol{\Phi}_1}{\partial \mathbf{x}_3} & \dfrac{\partial \boldsymbol{\Phi}_1}{\partial \mathbf{x}_4} \\ \dfrac{\partial \boldsymbol{\Phi}_2}{\partial \mathbf{x}_1} & \dfrac{\partial \boldsymbol{\Phi}_2}{\partial \mathbf{x}_2} & \dfrac{\partial \boldsymbol{\Phi}_2}{\partial \mathbf{x}_3} & \dfrac{\partial \boldsymbol{\Phi}_2}{\partial \mathbf{x}_4} \\ \dfrac{\partial \boldsymbol{\Phi}_3}{\partial \mathbf{x}_1} & \dfrac{\partial \boldsymbol{\Phi}_3}{\partial \mathbf{x}_2} & \dfrac{\partial \boldsymbol{\Phi}_3}{\partial \mathbf{x}_3} & \dfrac{\partial \boldsymbol{\Phi}_3}{\partial \mathbf{x}_4} \\ \dfrac{\partial \boldsymbol{\Phi}_4}{\partial \mathbf{x}_1} & \dfrac{\partial \boldsymbol{\Phi}_4}{\partial \mathbf{x}_2} & \dfrac{\partial \boldsymbol{\Phi}_4}{\partial \mathbf{x}_3} & \dfrac{\partial \boldsymbol{\Phi}_4}{\partial \mathbf{x}_4} \end{bmatrix}_{2n \times 2n} \begin{bmatrix} \mathbf{x}_2 \\ \mathbf{f}_0(\mathbf{x}) + \mathbf{G}_0(\mathbf{x})\mathbf{u} \\ \mathbf{x}_4 \\ \mathbf{u} \end{bmatrix}_{2n \times 1} \tag{29}$$

Therefore the component form of (29) can be expressed as

$$\mathbf{f}_1(\mathbf{x}) = \frac{\partial \boldsymbol{\Phi}_1}{\partial \mathbf{x}_1} \mathbf{x}_2 + \frac{\partial \boldsymbol{\Phi}_1}{\partial \mathbf{x}_2} \mathbf{f}_0(\mathbf{x}) + \frac{\partial \boldsymbol{\Phi}_1}{\partial \mathbf{x}_3} \mathbf{x}_4 + \left[\frac{\partial \boldsymbol{\Phi}_1}{\partial \mathbf{x}_2} \mathbf{G}_0(\mathbf{x}) + \frac{\partial \boldsymbol{\Phi}_1}{\partial \mathbf{x}_4} \right] \mathbf{u} \quad (30a)$$

$$\mathbf{f}_2(\mathbf{x}) = \frac{\partial \boldsymbol{\Phi}_2}{\partial \mathbf{x}_1} \mathbf{x}_2 + \frac{\partial \boldsymbol{\Phi}_2}{\partial \mathbf{x}_2} \mathbf{f}_0(\mathbf{x}) + \frac{\partial \boldsymbol{\Phi}_2}{\partial \mathbf{x}_3} \mathbf{x}_4 + \left[\frac{\partial \boldsymbol{\Phi}_2}{\partial \mathbf{x}_2} \mathbf{G}_0(\mathbf{x}) + \frac{\partial \boldsymbol{\Phi}_2}{\partial \mathbf{x}_4} \right] \mathbf{u} \quad (30b)$$

$$\mathbf{z}_4 = \frac{\partial \boldsymbol{\Phi}_3}{\partial \mathbf{x}_1} \mathbf{x}_2 + \frac{\partial \boldsymbol{\Phi}_3}{\partial \mathbf{x}_2} \mathbf{f}_0(\mathbf{x}) + \frac{\partial \boldsymbol{\Phi}_3}{\partial \mathbf{x}_3} \mathbf{x}_4 + \left[\frac{\partial \boldsymbol{\Phi}_3}{\partial \mathbf{x}_2} \mathbf{G}_0(\mathbf{x}) + \frac{\partial \boldsymbol{\Phi}_3}{\partial \mathbf{x}_4} \right] \mathbf{u} \quad (30c)$$

$$\mathbf{u} = \frac{\partial \boldsymbol{\Phi}_4}{\partial \mathbf{x}_1} \mathbf{x}_2 + \frac{\partial \boldsymbol{\Phi}_4}{\partial \mathbf{x}_2} \mathbf{f}_0(\mathbf{x}) + \frac{\partial \boldsymbol{\Phi}_4}{\partial \mathbf{x}_3} \mathbf{x}_4 + \left[\frac{\partial \boldsymbol{\Phi}_4}{\partial \mathbf{x}_2} \mathbf{G}_0(\mathbf{x}) + \frac{\partial \boldsymbol{\Phi}_4}{\partial \mathbf{x}_4} \right] \mathbf{u} \quad (30d)$$

Since the left hand sides of (30a), (30b) and (30c) contain no $\mathbf{u}$, so the corresponding terms in their respectively right hand side will be zero. On the other hand, the coefficient of $\mathbf{u}$ in (30d) should be $\mathbf{I}_m$. Hence, we have

$$\frac{\partial \boldsymbol{\Phi}_i}{\partial \mathbf{x}_2} \mathbf{G}_0(\mathbf{x}) + \frac{\partial \boldsymbol{\Phi}_i}{\partial \mathbf{x}_4} = \mathbf{0} \quad i = 1,2,3 \quad (31a)$$

$$\frac{\partial \boldsymbol{\Phi}_4}{\partial \mathbf{x}_2} \mathbf{G}_0(\mathbf{x}) + \frac{\partial \boldsymbol{\Phi}_4}{\partial \mathbf{x}_4} = \mathbf{I}_m \quad (31b)$$

Equations in (31a) form a set of partial differential equations whose solutions are $\boldsymbol{\Phi}_1$, $\boldsymbol{\Phi}_2$ and $\boldsymbol{\Phi}_3$. To ensure their solvability, we need to apply the famous Frobenius Theorem later. Equation (31b) can be solved at this moment by selecting

$$\boldsymbol{\Phi}_4 = \mathbf{x}_4 \quad (32)$$

and (30d) is reduced to $0 = \dfrac{\partial \boldsymbol{\Phi}_4}{\partial \mathbf{x}_1} \mathbf{x}_2 + \dfrac{\partial \boldsymbol{\Phi}_4}{\partial \mathbf{x}_2} \mathbf{f}_0(\mathbf{x}) + \dfrac{\partial \boldsymbol{\Phi}_4}{\partial \mathbf{x}_3} \mathbf{x}_4$. This partial differential equation is automatically satisfied with the selection of $\boldsymbol{\Phi}_4$ in (32). With the selections in (31a), we may reduce equations (30a), (30b) and (30c) to

$$\mathbf{f}_1(\mathbf{x}) = \frac{\partial \mathbf{\Phi}_1}{\partial \mathbf{x}_1}\mathbf{x}_2 + \frac{\partial \mathbf{\Phi}_1}{\partial \mathbf{x}_2}\mathbf{f}_0(\mathbf{x}) + \frac{\partial \mathbf{\Phi}_1}{\partial \mathbf{x}_3}\mathbf{x}_4 \tag{33a}$$

$$\mathbf{f}_2(\mathbf{x}) = \frac{\partial \mathbf{\Phi}_2}{\partial \mathbf{x}_1}\mathbf{x}_2 + \frac{\partial \mathbf{\Phi}_2}{\partial \mathbf{x}_2}\mathbf{f}_0(\mathbf{x}) + \frac{\partial \mathbf{\Phi}_2}{\partial \mathbf{x}_3}\mathbf{x}_4 \tag{33b}$$

$$\mathbf{z}_4 = \frac{\partial \mathbf{\Phi}_3}{\partial \mathbf{x}_1}\mathbf{x}_2 + \frac{\partial \mathbf{\Phi}_3}{\partial \mathbf{x}_2}\mathbf{f}_0(\mathbf{x}) + \frac{\partial \mathbf{\Phi}_3}{\partial \mathbf{x}_3}\mathbf{x}_4 \tag{33c}$$

If we can obtain $\mathbf{\Phi}_1$, $\mathbf{\Phi}_2$ and $\mathbf{\Phi}_3$ from (31a), then all terms in (26) can be computed by (33a), (33b) and (33c). So, now the problem becomes the solvability of the set of partial differential equations in (31a). Let us rewrite (31a) into the form

$$\begin{bmatrix} \dfrac{\partial \mathbf{\Phi}_i}{\partial \mathbf{x}_2} & \dfrac{\partial \mathbf{\Phi}_i}{\partial \mathbf{x}_4} \end{bmatrix}\begin{bmatrix} \mathbf{G}_0(\mathbf{x}) \\ \mathbf{I}_m \end{bmatrix} = \mathbf{0} \quad i = 1,2,3 \tag{34}$$

Hence, by the Frobenius Theorem, (31a) is solvable if the set of all columns of $\mathbf{G}(\mathbf{x}) = \begin{bmatrix} \mathbf{G}_0(\mathbf{x}) \\ \mathbf{I}_m \end{bmatrix}$ forms an involutive distribution. To check involutivity of this distribution, we compute the Lie bracket of the *i*th and *j*th columns of $\mathbf{G}$ as

$$[\mathbf{g}_i(\mathbf{q}) \quad \mathbf{g}_j(\mathbf{q})] = \begin{bmatrix} \dfrac{\partial \mathbf{g}_{0j}(\mathbf{q})}{\partial \mathbf{q}_u} & \dfrac{\partial \mathbf{g}_{0j}(\mathbf{q})}{\partial \mathbf{q}_a} \\ \dfrac{\partial \mathbf{e}_j}{\partial \mathbf{q}_u} & \dfrac{\partial \mathbf{e}_j}{\partial \mathbf{q}_a} \end{bmatrix}\begin{bmatrix} \mathbf{g}_{0i}(\mathbf{q}) \\ \mathbf{e}_i \end{bmatrix} - \begin{bmatrix} \dfrac{\partial \mathbf{g}_{0i}(\mathbf{q})}{\partial \mathbf{q}_u} & \dfrac{\partial \mathbf{g}_{0i}(\mathbf{q})}{\partial \mathbf{q}_a} \\ \dfrac{\partial \mathbf{e}_i}{\partial \mathbf{q}_u} & \dfrac{\partial \mathbf{e}_i}{\partial \mathbf{q}_a} \end{bmatrix}\begin{bmatrix} \mathbf{g}_{0j}(\mathbf{q}) \\ \mathbf{e}_j \end{bmatrix}$$

where $\mathbf{g}_i$ and $\mathbf{g}_{0i}$ are respectively the *i*th column of matrix $\mathbf{G}$ and $\mathbf{G}_0$, and $\mathbf{e}_i$ is the *i*-th basis vector. This can be rewritten to be

$$[\mathbf{g}_i(\mathbf{q}) \quad \mathbf{g}_j(\mathbf{q})] = \begin{bmatrix} \dfrac{\partial \mathbf{g}_{0j}(\mathbf{q})}{\partial \mathbf{q}_u}\mathbf{g}_{0i}(\mathbf{q}) + \dfrac{\partial \mathbf{g}_{0j}(\mathbf{q})}{\partial \mathbf{q}_a}\mathbf{e}_i - \dfrac{\partial \mathbf{g}_{0i}(\mathbf{q})}{\partial \mathbf{q}_u}\mathbf{g}_{0j}(\mathbf{q}) - \dfrac{\partial \mathbf{g}_{0i}(\mathbf{q})}{\partial \mathbf{q}_a}\mathbf{e}_j \\ \mathbf{0}_m \end{bmatrix}$$

Note that here we represent these column vectors with the dependence of the coordinate system $\mathbf{q}$ instead of $\mathbf{x}$ for convenience in computation. It is seen that if

$$\frac{\partial \mathbf{g}_{0j}(\mathbf{q})}{\partial \mathbf{q}_u}\mathbf{g}_{0i}(\mathbf{q}) + \frac{\partial \mathbf{g}_{0j}(\mathbf{q})}{\partial \mathbf{q}_a}\mathbf{e}_i - \frac{\partial \mathbf{g}_{0i}(\mathbf{q})}{\partial \mathbf{q}_u}\mathbf{g}_{0j}(\mathbf{q}) - \frac{\partial \mathbf{g}_{0i}(\mathbf{q})}{\partial \mathbf{q}_a}\mathbf{e}_j = \mathbf{0}$$

$\forall i, j = 1,...,m$, then the involutivity condition is satisfied and the set of partial differential equations in (31a) is solvable.

In summary, we have found the conditions to be satisfied to ensure transformation of (24) into (26) where the control input is effectively decoupled. As long as the system in (26) is in a cascade form, a stabilizing controller may be easier to find.

3.5 Decoupling for 2-DOF Systems

The coordinate transformation introduced in the last section is for the decoupling of a multi-input high-order underactuated system. Similar to the traditional input-to-state feedback linearization, although the set of partial differential equations in (31a) can be proved to be solvable, there is no general ways to solve them. In this section, we are going to investigate the decoupling of 2-DOF underactuated systems. Due to their relatively simple forms, some direct computations can be found to obtain the transformation without solving the set of partial differential equations. The class of 2-DOF underactuated systems is important because most of the benchmark systems fall into this category, e.g., pendubot, cart pole system, and TORA. In the following, we study the transformation of systems with dynamic models after the partial feedback linearization. Then, the transformation is given directly to a system model without going through the linearization.

Decoupling for Models after Partial Feedback Linearization

Consider a 2-DOF underactuated system represented in the form of (24) which is a model after collocated partial feedback linearization.

$$\dot{x}_1 = x_2$$
$$\dot{x}_2 = f_0(\mathbf{x}) + g_0(\mathbf{x})u$$
$$\dot{x}_3 = x_4 \tag{35}$$
$$\dot{x}_4 = u$$

where $\mathbf{x} = [x_1 \quad x_2 \quad x_3 \quad x_4]^T$ is the state vector. For simplicity, we would like to say that it is in the X-space. We can see that the same $u \in \Re$ is shared by the two subsystems. Let us apply the transformation (Olfati and Megretasi, 1998; Olfati, 2000, 2001, 2002)

$$z_1 = x_1 - \int_0^{x_3} g_0(s)ds$$
$$z_2 = x_2 - g_0(\mathbf{x})x_4$$
$$z_3 = x_3 \tag{36}$$
$$z_4 = x_4$$

Then, the dynamics of (35) in the space represented by the state vector $\mathbf{z} = [z_1 \quad z_2 \quad z_3 \quad z_4]^T$ is found by taking time derivative of (36) as

$$\dot{z}_1 = \dot{x}_1 - \frac{d}{dt}\int_0^{x_3} g_0(s)ds$$
$$= x_2 - \frac{d}{dt}\int_0^{z_3} g_0(s)ds$$
$$= [z_2 + g_0(\mathbf{x})x_4] - \frac{d}{dt}\int_0^{z_3} g_0(s)ds$$
$$= z_2 + g_0(\mathbf{z})z_4 - \frac{d}{dt}\int_0^{z_3} g_0(s)ds$$

$$\dot{z}_2 = \dot{x}_2 - \frac{dg_0(\mathbf{x})}{dt}x_4 - g_0(\mathbf{x})\dot{x}_4$$
$$= f_0(\mathbf{x}) + g_0(\mathbf{x})u - \frac{dg_0(\mathbf{x})}{dt}x_4 - g_0(\mathbf{x})u$$
$$= f_0(\mathbf{z}) - \frac{dg_0(\mathbf{z})}{dt}z_4$$

$$\dot{z}_3 = \dot{x}_3 = x_4 = z_4$$

$$\dot{z}_4 = \dot{x}_4 = u$$

These equations can be collected to give

$$\dot{z}_1 = z_2 + g_0(\mathbf{z})z_4 - \frac{d}{dt}\int_0^{z_3} g_0(s)ds$$

$$\dot{z}_2 = f_0(\mathbf{z}) - \frac{dg_0(\mathbf{z})}{dt}z_4 \tag{37}$$

$$\dot{z}_3 = z_4$$

$$\dot{z}_4 = u$$

We call (37) as the dynamic equation of (36) represented in the Z-space. We find that there is no control signal u in the first two lines in (37); therefore, the decoupling is effective. However, we see that the right hand sides of the first two lines in (37) are quite complex, and (37) does not satisfy the pure feedback condition. This implies that the backstepping procedure is not feasible here. We have to find some other control strategies for the system in this special cascade form.

Decoupling for Models without Feedback Linearization

Underactuated systems cannot be exactly feedback linearized. In the previous subsection we consider the decoupling of 2-DOF underactuated systems after partial feedback linearization where the actuated subsystem is a double integrator. However, actually the collocated feedback linearization is not really necessary for decoupling of an underactuated system. In this subsection, we are going to decouple a 2-DOF underactuated system in a model without using feedback linearization.

Consider a 2-DOF underactuated system represented in the form of (16) as

$$m_{11}(\mathbf{q})\ddot{q}_u + m_{12}(\mathbf{q})\ddot{q}_a + h_1(\mathbf{q},\dot{\mathbf{q}}) = 0 \tag{38a}$$

$$m_{21}(\mathbf{q})\ddot{q}_u + m_{22}(\mathbf{q})\ddot{q}_a + h_2(\mathbf{q},\dot{\mathbf{q}}) = u_2 \tag{38b}$$

where $\mathbf{q} = [q_u \quad q_a]^T \in \Re^2$, and

$$h_1(\mathbf{q},\dot{\mathbf{q}}) = c_{11}(\mathbf{q},\dot{\mathbf{q}})\dot{q}_u + c_{12}(\mathbf{q},\dot{\mathbf{q}})\dot{q}_a + g_1(\mathbf{q})$$

$$h_2(\mathbf{q},\dot{\mathbf{q}}) = c_{21}(\mathbf{q},\dot{\mathbf{q}})\dot{q}_u + c_{22}(\mathbf{q},\dot{\mathbf{q}})\dot{q}_a + g_2(\mathbf{q})$$

From (38a), we may find

$$\ddot{q}_u = -m_{11}^{-1}(\mathbf{q})m_{12}(\mathbf{q})\ddot{q}_a - m_{11}^{-1}(\mathbf{q})h_1(\mathbf{q},\dot{\mathbf{q}}) \qquad (39)$$

Substitute (39) into (38b) to give

$$m_{21}(-m_{11}^{-1}m_{12}\ddot{q}_a - m_{11}^{-1}h_1) + m_{22}\ddot{q}_a + h_2 = u_2$$

After some straightforward arrangement, we get

$$\ddot{q}_a = \underbrace{(m_{22} - m_{21}m_{11}^{-1}m_{12})^{-1}[m_{21}m_{11}^{-1}h_1(\mathbf{q},\dot{\mathbf{q}}) - h_2(\mathbf{q},\dot{\mathbf{q}})]}_{f_4(\mathbf{q},\dot{\mathbf{q}})} + \underbrace{(m_{22} - m_{21}m_{11}^{-1}m_{12})^{-1}}_{b_4(\mathbf{q})}u_2$$

This can be represented as

$$\ddot{q}_a = f_4(\mathbf{q},\dot{\mathbf{q}}) + b_4(\mathbf{q})u_2 \qquad (40)$$

where

$$f_4(\mathbf{q},\dot{\mathbf{q}}) = (m_{22} - m_{21}m_{11}^{-1}m_{12})^{-1}[m_{21}m_{11}^{-1}h_1(\mathbf{q},\dot{\mathbf{q}}) - h_2(\mathbf{q},\dot{\mathbf{q}})]$$

$$b_4(\mathbf{q}) = (m_{22} - m_{21}m_{11}^{-1}m_{12})^{-1}$$

Next, plugging (40) into (39) to have

$$\ddot{q}_u = -m_{11}^{-1}m_{12}[f_4(\mathbf{q},\dot{\mathbf{q}}) + b_4(\mathbf{q})u_2] - m_{11}^{-1}h_1(\mathbf{q},\dot{\mathbf{q}})$$
$$= \underbrace{[-m_{11}^{-1}m_{12}f_4(\mathbf{q},\dot{\mathbf{q}}) - m_{11}^{-1}h_1(\mathbf{q},\dot{\mathbf{q}})]}_{f_2(\mathbf{q},\dot{\mathbf{q}})} + \underbrace{[-m_{11}^{-1}m_{12}b_4(\mathbf{q})]}_{b_2(\mathbf{q})}u_2$$

Again, this can be expressed in a simpler form

$$\ddot{q}_u = f_2(\mathbf{q},\dot{\mathbf{q}}) + b_2(\mathbf{q})u_2 \qquad (41)$$

where $f_2(\mathbf{q},\dot{\mathbf{q}}) = -m_{11}^{-1}m_{12}f_4(\mathbf{q},\dot{\mathbf{q}}) - m_{11}^{-1}h_1(\mathbf{q},\dot{\mathbf{q}})$ and $b_2(\mathbf{q}) = -m_{11}^{-1}m_{12}b_4(\mathbf{q})$.
Collect (40) and (41) to have the representation of the system as

$$\ddot{q}_u = f_2(\mathbf{q},\dot{\mathbf{q}}) + b_2(\mathbf{q})u \qquad (42a)$$

$$\ddot{q}_a = f_4(\mathbf{q},\dot{\mathbf{q}}) + b_4(\mathbf{q})u \qquad (42b)$$

where $u = u_2$. Define the state vector

$$\mathbf{x} = [x_1 \quad x_2 \quad x_3 \quad x_4]^T = [q_u \quad \dot{q}_u \quad q_a \quad \dot{q}_a]^T$$

and we have the system equation in the X-space as

$$
\begin{aligned}
\dot{x}_1 &= x_2 \\
\dot{x}_2 &= f_2(\mathbf{x}) + b_2(\mathbf{x})u \\
\dot{x}_3 &= x_4 \\
\dot{x}_4 &= f_4(\mathbf{x}) + b_4(\mathbf{x})u
\end{aligned}
\qquad (43)
$$

Note that the representation (43) is different from the one in (35). Firstly, the actuated subsystem in (35) is a double integrator while the one in (43) contains nonlinearities. Secondly, there is no coefficient in front of the control signal of the actuated system in (35), but there is $b_4(\mathbf{x}) = (m_{22} - m_{21}m_{11}^{-1}m_{12})^{-1}$ in (43) as the coefficient of the control of the actuated system. Therefore, we may not use (36) to decouple (43) directly. Let us now consider the transformation

$$z_1 = x_1 - \int_0^{x_3} \frac{b_2(s)}{b_4(s)}\,ds$$

$$z_2 = x_2 - \frac{b_2(\mathbf{x})}{b_4(\mathbf{x})}x_4 \qquad (44)$$

$$z_3 = x_3$$

$$z_4 = x_4$$

Then, the dynamics of (43) represented in the Z-space is found by taking time derivative of (44) as

$$\dot{z}_1 = \dot{x}_1 - \frac{d}{dt}\int_0^{x_3}\frac{b_2(s)}{b_4(s)}ds$$

$$= x_2 - \frac{d}{dt}\int_0^{z_3}\frac{b_2(s)}{b_4(s)}ds$$

$$= z_2 + \frac{b_2(\mathbf{z})}{b_4(\mathbf{z})}z_4 - \frac{d}{dt}\int_0^{z_3}\frac{b_2(s)}{b_4(s)}ds$$

$$\dot{z}_2 = \dot{x}_2 - \frac{d}{dt}\left[\frac{b_2}{b_4}\right]x_4 - \frac{b_2}{b_4}\dot{x}_4$$

$$= f_2(\mathbf{x}) + b_2(\mathbf{x})u - \frac{d}{dt}\left[\frac{b_2}{b_4}\right]x_4 - \frac{b_2}{b_4}[f_4(\mathbf{x}) + b_4(\mathbf{x})u]$$

$$= f_2(\mathbf{z}) - \frac{b_2(\mathbf{z})}{b_4(\mathbf{z})}f_4(\mathbf{z}) - \frac{d}{dt}\left[\frac{b_2(\mathbf{z})}{b_4(\mathbf{z})}\right]z_4$$

$$\dot{z}_3 = \dot{x}_3 = x_4 = z_4$$

$$\dot{z}_4 = \dot{x}_4 = f_4(\mathbf{z}) + b_4(\mathbf{z})u$$

These equations can be collected to give

$$\dot{z}_1 = z_2 + \frac{b_2(\mathbf{z})}{b_4(\mathbf{z})}z_4 - \frac{d}{dt}\int_0^{z_3}\frac{b_2(s)}{b_4(s)}ds$$

$$\dot{z}_2 = f_2(\mathbf{z}) - \frac{b_2(\mathbf{z})}{b_4(\mathbf{z})}f_4(\mathbf{z}) - \frac{d}{dt}\left[\frac{b_2(\mathbf{z})}{b_4(\mathbf{z})}\right]z_4 \qquad (45)$$

$$\dot{z}_3 = z_4$$

$$\dot{z}_4 = f_4(\mathbf{z}) + b_4(\mathbf{z})u$$

This is the system dynamics in the Z-space where only one u appears in the equation. It is seen that the actuated subsystem in (45) is the same as the one in (43), but the unactuated system is rather complex. Let us define

$$d_1(\mathbf{z}) = \frac{b_2(\mathbf{z})}{b_4(\mathbf{z})} z_4 - \frac{d}{dt} \int_0^{z_3} \frac{b_2(s)}{b_4(s)} ds$$

$$d_2(\mathbf{z}) = f_2(\mathbf{z}) - \frac{b_2(\mathbf{z})}{b_4(\mathbf{z})} f_4(\mathbf{z}) - \frac{d}{dt} \left[\frac{b_2(\mathbf{z})}{b_4(\mathbf{z})} \right] z_4 - z_3 \qquad (46)$$

$$d_3(\mathbf{z}) = 0$$

$$d_4(\mathbf{z}) = f_4(\mathbf{z})$$

then (45) is able to be expressed as

$$\begin{aligned}
\dot{z}_1 &= z_2 + d_1(\mathbf{z}) \\
\dot{z}_2 &= z_3 + d_2(\mathbf{z}) \\
\dot{z}_3 &= z_4 + d_3(\mathbf{z}) \\
\dot{z}_4 &= d_4(\mathbf{z}) + b_4(\mathbf{z})u
\end{aligned} \qquad (47)$$

It is seen that (47) is in a special cascade form with only one control input appears in the last subsystem. It may be possible to construct a controller to stabilize the system in the Z-space, so that stabilization of the system in the X-space follows.

Since $d_1(\mathbf{z})$ and $d_2(\mathbf{z})$ in (46) could be so complex that their actual forms are very difficult to compute. A reasonable way is to regard them as uncertainties in (47). Because they enter the system in a mismatched fashion, not so many control strategies are feasible. The backstepping procedure is well-known to be effective in dealing with mismatched uncertainties. However, because $d_1(\mathbf{z})$ and $d_2(\mathbf{z})$ are functions of the states, equation (47) does not fit in the pure feedback form. This implies that the backstepping procedure is not applicable to this system. In addition, both $d_1(\mathbf{z})$ and $d_2(\mathbf{z})$ are not always linearly parameterizable into the multiplication of known regressors and unknown constant parameter vectors. Hence, conventional adaptive strategies are not able to give proper update laws to ensure closed-loop stability. On the other hand, it is not practical to assume that the variation bounds for $d_1(\mathbf{z})$ and $d_2(\mathbf{z})$ are available for all cases; therefore, the robust designs fail in general.

In summary, a 2-DOF underactuated system can be represented in the special cascade structure in (47) which does not satisfy the standard pure feedback form and the backstepping is not feasible. The mismatched uncertainties $d_1(\mathbf{z})$ and $d_2(\mathbf{z})$ cannot be covered by using traditional adaptive designs or robust schemes either. In the next chapter, we are going to design a multiple-surface sliding controller to cope with the mismatched uncertainties incorporated with the function approximated techniques to update the uncertainties online.

3.6 Decoupling of Single-Input High-Order Systems

In the previous section, we investigate the decoupling of 2-DOF underactuated systems into a special cascade form without solving the set of partial differential equations. In this section, we would like to extend the same concept to high-order systems. The basic idea comes from the fact that each mechanical subsystem in an underactuated system can be modeled as a second order system whose control input can thus be decoupled using similar transformations in (44). In the following, we clarify this concept on a 3-DOF system first followed by the introduction of a general formulation.

Decoupling of 3-DOF Systems

Consider a 3-DOF underactuated system represented in a 6th order equation

$$
\begin{aligned}
\dot{x}_1 &= x_2 \\
\dot{x}_2 &= f_2(\mathbf{x}) + b_2(\mathbf{x})u \\
\dot{x}_3 &= x_4 \\
\dot{x}_4 &= f_4(\mathbf{x}) + b_4(\mathbf{x})u \\
\dot{x}_5 &= x_6 \\
\dot{x}_6 &= f_6(\mathbf{x}) + b_6(\mathbf{x})u
\end{aligned}
\tag{48}
$$

where the control u is shared by the 3 subsystems, so it is underactuated. Now let us consider the coordinate transformation similar to (44) as

$$z_1 = x_1 - \int_0^{x_3} \frac{b_2(s)}{b_4(s)} ds$$

$$z_2 = x_2 - \frac{b_2(\mathbf{x})}{b_4(\mathbf{x})} x_4$$

$$z_3 = x_3 - \int_0^{x_5} \frac{b_4(s)}{b_6(s)} ds \qquad (49)$$

$$z_4 = x_4 - \frac{b_4(\mathbf{x})}{b_6(\mathbf{x})} x_4$$

$$z_5 = x_5$$

$$z_6 = x_6$$

Then, the dynamics of (48) represented in the Z-space is found by taking time derivative of (49) as

$$\dot{z}_1 = z_2 + \frac{b_2(\mathbf{z})}{b_4(\mathbf{z})} z_4 - \frac{d}{dt} \int_0^{z_3} \frac{b_2(s)}{b_4(s)} ds$$

$$\dot{z}_2 = f_2(\mathbf{z}) - \frac{b_2(\mathbf{z})}{b_4(\mathbf{z})} f_4(\mathbf{z}) - \frac{d}{dt}\left[\frac{b_2(\mathbf{z})}{b_4(\mathbf{z})}\right] z_4$$

$$\dot{z}_3 = z_4 + \frac{b_4(\mathbf{z})}{b_6(\mathbf{z})} z_6 - \frac{d}{dt} \int_0^{z_5} \frac{b_4(s)}{b_6(s)} ds \qquad (50)$$

$$\dot{z}_4 = f_4(\mathbf{z}) - \frac{b_4(\mathbf{z})}{b_6(\mathbf{z})} f_6(\mathbf{z}) - \frac{d}{dt}\left[\frac{b_4(\mathbf{z})}{b_6(\mathbf{z})}\right] z_6$$

$$\dot{z}_5 = z_6$$

$$\dot{z}_6 = f_6(\mathbf{z}) + b_6(\mathbf{z})u$$

This is the system dynamics in the Z-space where only one u appears in the equation so it is in a special cascade form. This implies that the concept in the transformation (44) can be used in decoupling high order underactuated systems. In the next subsection, we extend the method to an n-th order system.

Decoupling of *n*-th Order Systems

Let us consider an *n*-th order underactuated system represented in the state space as

$$
\begin{aligned}
\dot{x}_1 &= x_2 \\
\dot{x}_2 &= f_2(\mathbf{x}) + b_2(\mathbf{x})u \\
\dot{x}_3 &= x_4 \\
\dot{x}_4 &= f_4(\mathbf{x}) + b_4(\mathbf{x})u \\
&\vdots \\
\dot{x}_{n-1} &= x_n \\
\dot{x}_n &= f_n(\mathbf{x}) + b_n(\mathbf{x})u
\end{aligned}
\tag{51}
$$

where *n* is an even number. The coordinate transformation is defined as

$$
\begin{aligned}
z_1 &= x_1 - \int_0^{x_3} \frac{b_2(s)}{b_4(s)} ds \\
z_2 &= x_2 - \frac{b_2(\mathbf{x})}{b_4(\mathbf{x})} x_4 \\
z_3 &= x_3 - \int_0^{x_5} \frac{b_4(s)}{b_6(s)} ds \\
z_4 &= x_4 - \frac{b_4(\mathbf{x})}{b_6(\mathbf{x})} x_4 \\
&\vdots \\
z_{n-3} &= x_{n-3} - \int_0^{x_{n-1}} \frac{b_{n-2}(s)}{b_n(s)} ds \\
z_{n-2} &= x_{n-2} - \frac{b_{n-2}(\mathbf{x})}{b_n(\mathbf{x})} x_n \\
z_{n-1} &= x_{n-1} \\
z_n &= x_n
\end{aligned}
\tag{52}
$$

Hence, the system dynamics in the Z-space can be found as

$$\dot{z}_1 = z_2 + \frac{b_2(\mathbf{z})}{b_4(\mathbf{z})} z_4 - \frac{d}{dt} \int_0^{z_3} \frac{b_2(s)}{b_4(s)} ds$$

$$\dot{z}_2 = f_2(\mathbf{z}) - \frac{b_2(\mathbf{z})}{b_4(\mathbf{z})} f_4(\mathbf{z}) - \frac{d}{dt} \left[\frac{b_2(\mathbf{z})}{b_4(\mathbf{z})} \right] z_4$$

$$\dot{z}_3 = z_4 + \frac{b_4(\mathbf{z})}{b_6(\mathbf{z})} z_6 - \frac{d}{dt} \int_0^{z_5} \frac{b_4(s)}{b_6(s)} ds$$

$$\dot{z}_4 = f_4(\mathbf{z}) - \frac{b_4(\mathbf{z})}{b_6(\mathbf{z})} f_6(\mathbf{z}) - \frac{d}{dt} \left[\frac{b_4(\mathbf{z})}{b_6(\mathbf{z})} \right] z_6$$

$$\vdots \tag{53}$$

$$\dot{z}_{n-3} = z_{n-2} + \frac{b_{n-2}(\mathbf{z})}{b_n(\mathbf{z})} z_n - \frac{d}{dt} \int_0^{z_{n-1}} \frac{b_{n-2}(s)}{b_n(s)} ds$$

$$\dot{z}_{n-2} = f_{n-2}(\mathbf{z}) - \frac{b_{n-2}(\mathbf{z})}{b_n(\mathbf{z})} f_n(\mathbf{z}) - \frac{d}{dt} \left[\frac{b_{n-2}(\mathbf{z})}{b_n(\mathbf{z})} \right] z_n$$

$$\dot{z}_{n-1} = z_n$$

$$\dot{z}_n = f_n(\mathbf{z}) + b_n(\mathbf{z})u$$

Now, let us define

$$d_1(\mathbf{z}) = \frac{b_2(\mathbf{z})}{b_4(\mathbf{z})} z_4 - \frac{d}{dt} \int_0^{z_3} \frac{b_2(s)}{b_4(s)} ds$$

$$d_2(\mathbf{z}) = f_2(\mathbf{z}) - \frac{b_2(\mathbf{z})}{b_4(\mathbf{z})} f_4(\mathbf{z}) - \frac{d}{dt} \left[\frac{b_2(\mathbf{z})}{b_4(\mathbf{z})} \right] z_4 - z_3$$

$$\vdots$$

$$d_{n-3}(\mathbf{z}) = \frac{b_{n-2}(\mathbf{z})}{b_n(\mathbf{z})} z_n - \frac{d}{dt} \int_0^{z_{n-1}} \frac{b_{n-2}(s)}{b_n(s)} ds \tag{54}$$

$$d_{n-2}(\mathbf{z}) = f_{n-2}(\mathbf{z}) - \frac{b_{n-2}(\mathbf{z})}{b_n(\mathbf{z})} f_n(\mathbf{z}) - \frac{d}{dt} \left[\frac{b_{n-2}(\mathbf{z})}{b_n(\mathbf{z})} \right] z_n - z_{n-1}$$

$$d_{n-1}(\mathbf{z}) = 0$$

$$d_n(\mathbf{z}) = f_n(\mathbf{z})$$

Then, (53) can be written in a standard form similar to (47) as

$$\dot{z}_1 = z_2 + d_1(\mathbf{z})$$
$$\dot{z}_2 = z_3 + d_2(\mathbf{z})$$
$$\vdots$$
$$\dot{z}_{n-3} = z_{n-2} + d_{n-3}(\mathbf{z}) \qquad (55)$$
$$\dot{z}_{n-2} = z_{n-1} + d_{n-2}(\mathbf{z})$$
$$\dot{z}_{n-1} = z_n + d_{n-1}(\mathbf{z})$$
$$\dot{z}_n = d_n(\mathbf{z}) + b_n(\mathbf{z})u$$

This is an n-th order cascade system with mismatched uncertainties. In the next chapter, an FAT-based adaptive multi-surface sliding controller is designed to stabilize the system in the Z-space which further ensure stability in the X-space.

Chapter 4

Controller Design

In this chapter, an adaptive multiple-surface sliding controller (AMSSC) based on the function approximation techniques will be derived to stabilize a class of uncertain underactuated systems. In section 4.1, we clarify the control problem by investigating the system models obtained in chapter 3 more closely. The admissible uncertainties are identified in the X-space and carried to the Z-space without disturbing the coordinate transformation. An AMSSC is given in section 4.2 for an n-th order system in the cascade form represented in the Z-space. It is proved that if the approximation error is ignorable, the asymptotic stability of the output error and boundedness of the internal signals can be obtained. When considering the approximation error, the output error is guaranteed to be uniformly ultimately bounded, and the transient response is bounded by an exponential function. A summary is given in section 4.3 to combine the results from chapter 3 to chapter 4.

4.1 Control Problem Formulation

In section 3.5, decoupling techniques for 2-DOF underactuated systems with or without the partial feedback linearization are introduced. In the following, we firstly show that the decoupling technique with the partial feedback linearization does not allow the system to contain uncertainties. On the other hand, we will also see that the one without the partial feedback linearization allows the system in the X-space to have uncertainties. To cover a wider class of systems, a control problem is formulated later for n-th order underactuated systems without using the partial feedback linearization.

Systems with Partial Feedback Linearization

The coordinate transformation introduced in subsection 3.5.1 can be summarized in Fig. 4.1 where the system in the X-space is represented by (3.35), the system in the Z-space is by (3.37) and the coordinate transformation is (3.36). It is noted that equation (3.36) involves the term $g_0 = m_{11}^{-1} m_{12}$, and hence g_0 in (3.35) may not contain uncertainties; otherwise the coordinate transformation is not realizable. Actually, f_0 in (3.35) may not contain uncertainties either. This is because in the partial feedback linearization we have to use the control (3.19) to linearize the actuated system into a double integrator. This implies that we need to know the term

$$\beta = (m_{22} - m_{21} m_{11}^{-1} m_{12})^{-1} (h_2 - m_{21} m_{11}^{-1} h_1)$$

which is a function of h_1 and h_2, and hence a function of the matrix **C** and vector **g**. In addition, the matrix **M** is also needed to be available to compute β, so this shows that the system in the X-space may not contain any uncertainty when the partial feedback linearization is preferable.

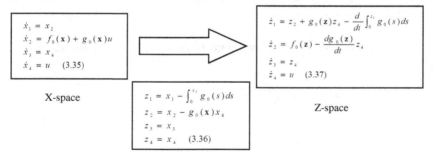

Figure 4.1. The decoupling of a 2-DOF system with the partial feedback linearization introduced in subsection 3.5.1 is summarized. Equation (3.35) is the system in the X-space and (3.37) represents the system in the Z-space. The coordinate transformation is shown in (3.36). It is seen that in the X-space the control u is shared by the two subsystem resulting in an underactuated dynamics, while in the Z-space there is only one u appears in the second subsystem showing an effective decoupling.

Systems without Partial Feedback Linearization

Now, let us see the transformation based on the system model without going through partial feedback linearization as shown in Fig. 4.2.

$$
\begin{aligned}
\dot{x}_1 &= x_2 \\
\dot{x}_2 &= f_2(\mathbf{x}) + b_2(\mathbf{x})u \\
\dot{x}_3 &= x_4 \qquad (3.43) \\
\dot{x}_4 &= f_4(\mathbf{x}) + b_4(\mathbf{x})u
\end{aligned}
$$

X-space

$$
\begin{aligned}
\dot{z}_1 &= z_2 + \frac{b_2(\mathbf{z})}{b_4(\mathbf{z})} z_4 - \frac{d}{dt} \int_0^{z_3} \frac{b_2(s)}{b_4(s)} ds \\
\dot{z}_2 &= f_2(\mathbf{z}) - \frac{b_2(\mathbf{z})}{b_4(\mathbf{z})} f_4(\mathbf{z}) - \frac{d}{dt}\left[\frac{b_2(\mathbf{z})}{b_4(\mathbf{z})}\right] z_4 \\
\dot{z}_3 &= z_4 \\
\dot{z}_4 &= f_4(\mathbf{z}) + b_4(\mathbf{z})u \qquad (3.45)
\end{aligned}
$$

Z-space

$$
\begin{aligned}
z_1 &= x_1 - \int_0^{x_3} \frac{b_2(s)}{b_4(s)} ds \\
z_2 &= x_2 - \frac{b_2(\mathbf{x})}{b_4(\mathbf{x})} x_4 \\
z_3 &= x_3 \\
z_4 &= x_4 \qquad (3.44)
\end{aligned}
$$

Figure 4.2. The decoupling of a 2-DOF system without partial feedback linearization introduced in subsection 3.5.2 is summarized. Equation (3.43) is the system in the X-space and (3.45) represents the system in the Z-space. The coordinate transformation is shown in (3.44). It is seen that in the X-space the control u is shared by the two subsystems resulting in an underactuated dynamics, while in the Z-space there is only one u appears in the second subsystem showing an effective decoupling.

It is seen from the coordinate transformation in equation (3.44) that both

$$
b_2(\mathbf{x}) = -m_{11}^{-1} m_{12}(m_{22} - m_{21} m_{11}^{-1} m_{12})^{-1} \quad \text{and}
$$

$$
b_4(\mathbf{x}) = (m_{22} - m_{21} m_{11}^{-1} m_{12})^{-1}
$$

are present, so they have to be known. This implies that the inertia matrix **M** is not able to contain uncertainties. However, both f_2 and f_4 in (3.43) are independent to the transformation; therefore, there is no any limitation on their availability. Since the precise forms of f_2 and f_4 are respectively

$$
f_2 = -m_{11}^{-1} m_{12}(m_{22} - m_{21} m_{11}^{-1} m_{12})^{-1}(m_{21} m_{11}^{-1} h_1 - h_2) - m_{11}^{-1} h_1
$$

$$f_4 = (m_{22} - m_{21}m_{11}^{-1}m_{12})^{-1}(m_{21}m_{11}^{-1}h_1 - h_2)$$

the functions h_1 and h_2 are allowed to have uncertainties. This is equivalent to say that the matrix $\mathbf{C}$ and vector $\mathbf{g}$ are able to be unknown. Therefore, in this book, we shall use the model without going through the partial feedback linearization to permit the system to contain uncertainties.

The Control Problem

Let us consider the n-th order underactuated system represented in the Z-space shown in (3.55) without going through the partial feedback linearization as

$$\dot{z}_1 = z_2 + d_1(\mathbf{z})$$
$$\dot{z}_2 = z_3 + d_2(\mathbf{z})$$
$$\vdots$$
$$\dot{z}_{n-3} = z_{n-2} + d_{n-3}(\mathbf{z}) \tag{1}$$
$$\dot{z}_{n-2} = z_{n-1} + d_{n-2}(\mathbf{z})$$
$$\dot{z}_{n-1} = z_n + d_{n-1}(\mathbf{z})$$
$$\dot{z}_n = d_n(\mathbf{z}) + b_n(\mathbf{z})u$$

where n is an even number, $d_i = \dfrac{b_{i+1}(\mathbf{z})}{b_{i+3}(\mathbf{z})}z_{i+3} - \dfrac{d}{dt}\displaystyle\int_0^{z_{i+2}} \dfrac{b_{i+1}(s)}{b_{i+3}(s)}ds$, for all

$i = 1,3,...,n-3$, $d_j = f_j(\mathbf{z}) - \dfrac{b_j(\mathbf{z})}{b_{j+2}(\mathbf{z})}f_{j+2}(\mathbf{z}) - \dfrac{d}{dt}\left[\dfrac{b_j(\mathbf{z})}{b_{j+2}(\mathbf{z})}\right]z_{j+2} - z_{j+1}$, for

all $j = 2,4,...,n-2$, $d_{n-1} = 0$ and $d_n = f_n$. Since $b_i(\mathbf{z})$, $i = 2,4,...,n$ have to be known to ensure feasibility of the coordinate transformation in (3.52), $d_i, \forall i = 1,3,...,n-3$ are known. On the contrary, since $f_i(\mathbf{z})$, $i = 2,4,...,n$ are unknown, $d_i, \forall i = 2,4,...,n-2$ are not known. In section

3.5, we have mentioned that the term $\dfrac{d}{dt}\displaystyle\int_0^{z_{i+2}} \dfrac{b_{i+1}(s)}{b_{i+3}(s)}ds$ in the

computation of $d_i, \forall i = 1,3,...,n-3$ could be very complex even though we know exact forms of b_{i+1} and b_{i+3}. So, we would like to assume that

$d_i, \forall i = 1,3,...,n-3$ are all unknown functions to avoid their computations which will greatly simplify the implementation of the control law. Hence, we may have a statement for the control problem as: Given a cascade system (1) in the Z-space where $d_i, \forall i = 1,2,...,n$ are all unknown functions but b_n is known, we are required to design a stabilizing controller to have good performance.

4.2 Controller Design

Since (1) is not in the pure feedback form, we may not apply the backstepping procedure to deal with the mismatched uncertainties $d_i, \forall i = 1,2,...,n-1$. The multiple-surface sliding controller introduced in section 2.10 cannot be used either, because the variation bounds of $d_i, \forall i = 1,2,...,n$ are not assumed to be known. This also implies that most of the robust designs fail here due to the unavailability of the variation bounds. In addition, we are not able to assume that $d_i, \forall i = 1,2,...,n$ are constants or they are linearly parameterizable into a multiplication of a known regressor and a vector of unknown constant parameters. Therefore, the traditional adaptive controllers are not feasible. In this section, we would like to use the function approximation technique to represent all uncertainties as a finite combination of orthogonal basis functions. The adaptive multiple-surface sliding controller is then designed so that proper update laws can be constructed to ensure stability of the closed loop system. Although we know b_n in (1), we are going to assume that it is a bounded uncertainty to show that AMSSC can be applied to more complex control problems.

Now let us consider the uncertain system in the Z-space represented in (1) where $\mathbf{z} = [z_1 \quad z_2 \quad \cdots \quad z_n]^T \in \Omega$ and Ω is a compact subset of $\mathfrak{R}^n$. The general uncertainties $d_i, \forall i = 1,2,...,n$ satisfy $z_{i+1} + d_i(\mathbf{z}) \neq 0$, and their variation bounds are not known. The uncertain function $b_n(\mathbf{z})$ is assumed to be bounded by $0 < b_{min} \leq b_n(\mathbf{z}) \leq b_{max}$ for some known functions b_{min} and b_{max} for all $\mathbf{z} \in \Omega$. Let $b_m = \sqrt{b_{min}b_{max}}$ be the nominal function of $b_n(\mathbf{z})$, and then we may represent $b_n(\mathbf{z})$ in the form $b_n(\mathbf{z}) = b_m(\mathbf{z})\Delta b(\mathbf{z})$ where $\Delta b(\mathbf{z})$ is the multiplicative uncertainty satisfying

$$0 < \beta_{min} \equiv \frac{b_{min}}{b_m} \leq \Delta b \leq \frac{b_{max}}{b_m} \equiv \beta_{max}$$

We would like to design a controller such that the system state $\mathbf{z}$ tracks the desired trajectory $\mathbf{z}_d \in \Omega_d$ where Ω_d is a compact subset of Ω.

The controller can be designed in several steps followed by stability analysis. Let us define the error signals

$$\begin{aligned} s_1 &= z_1 - z_{1d} \\ s_2 &= z_2 - z_{2d} \\ &\vdots \\ s_n &= z_n - z_{nd} \end{aligned} \tag{2}$$

where z_{id} is the desired trajectory for the ith state, $i = 1,...,n$. These error signals can be regarded as a set of error surfaces in some error space. This coins the name "multiple-surface control."

STEP 1: The dynamics of s_1 can be found by taking the time derivative of the first equation in (2) as

$$\begin{aligned} \dot{s}_1 &= \dot{z}_1 - \dot{z}_{1d} \\ &= z_2 + d_1 - \dot{z}_{1d} \\ &= s_2 + z_{2d} + d_1 - \dot{z}_{1d} \end{aligned} \tag{3}$$

To stabilize this subsystem, one possibility is to regard z_{2d} as a virtual control. With the selection

$$z_{2d} = \dot{z}_{1d} - \hat{d}_1 - c_1 s_1 \tag{4}$$

where $\hat{d}_1$ is an estimate of d_1, and c_1 is a positive constant, equation (3) becomes

$$\dot{s}_1 = -c_1 s_1 + s_2 + (d_1 - \hat{d}_1) \tag{5}$$

STEP 2: Let us derive the dynamics of s_2 as

$$\dot{s}_2 = s_3 + z_{3d} + d_2 - \dot{z}_{2d} \tag{6}$$

where $\dot{z}_{2d}$ can be computed from (4) with

$$\begin{aligned}
\dot{z}_{2d} &= \ddot{z}_{1d} - \dot{\hat{d}}_1 - c_1[-c_1 s_1 + s_2 + (d_1 - \hat{d}_1)] \\
&= \underbrace{\ddot{z}_{1d} - \dot{\hat{d}}_1 + c_1^2 s_1 - c_1 s_2 - c_1 \hat{d}_1}_{\dot{z}_{2dk}} + \underbrace{(-c_1 d_1)}_{\dot{z}_{2du}} \\
&= \dot{z}_{2dk} + \dot{z}_{2du}
\end{aligned} \tag{7}$$

where $\dot{z}_{2dk} = \ddot{z}_{1d} - \dot{\hat{d}}_1 + c_1^2 s_1 - c_1 s_2 - c_1 \hat{d}_1$ and $\dot{z}_{2du} = -c_1 d_1$ are respectively the known and unknown parts of $\dot{z}_{2d}$. Finally, (6) becomes

$$\begin{aligned}
\dot{s}_2 &= s_3 + z_{3d} + d_2 - \dot{z}_{2dk} - \dot{z}_{2du} \\
&= s_3 + z_{3d} + \overline{d}_2 - \dot{z}_{2dk}
\end{aligned} \tag{8}$$

where $\overline{d}_2 = d_2 - \dot{z}_{2du}$ is a lumped uncertainty. Now we may regard z_{3d} as a virtual control input to stabilized (8), and it is selected as

$$z_{3d} = \dot{z}_{2dk} - \hat{\overline{d}}_2 - c_2 s_2 \tag{9}$$

where $\hat{\overline{d}}_2$ is an estimate of $\overline{d}_2$, and c_2 is a positive constant. With this selection of z_{3d}, equation (8) can be rewritten as

$$\dot{s}_2 = -c_2 s_2 + s_3 + (\overline{d}_2 - \hat{\overline{d}}_2) \tag{10}$$

Up to this point, we know that each step will have a selection of the virtual control for the subsystem stabilization and a differential equation describing the error dynamics can be obtained. We may thus give the rules below for the *i*th subsystem.

STEP i: $i = 2,\ldots,n-1$

$$\dot{s}_i = -c_i s_i + s_{i+1} + (\overline{d}_i - \hat{\overline{d}}_i) \tag{11a}$$

$$z_{(i+1)d} = \dot{z}_{idk} - \hat{\overline{d}}_i - c_i s_i \tag{11b}$$

$$\begin{aligned}
\dot{z}_{id} &= \frac{d}{dt}(\dot{z}_{(i-1)dk} - \hat{\overline{d}}_{(i-1)} - c_{i-1}s_{i-1}) \\
&= \ddot{z}_{(i-1)dk} - \dot{\hat{\overline{d}}}_{i-1} - c_{i-1}[-c_{i-1}s_{i-1} + s_i + (\overline{d}_{i-1} - \hat{\overline{d}}_{i-1})] \\
&= \underbrace{\ddot{z}_{(i-1)dk} - \dot{\hat{\overline{d}}}_{i-1} - c_{i-1}(-c_{i-1}s_{i-1} + s_i - \hat{\overline{d}}_{i-1})}_{\dot{z}_{idk}} + \underbrace{(-c_{i-1}\overline{d}_{i-1})}_{\dot{z}_{idu}} \\
&= \dot{z}_{idk} + \dot{z}_{idu}
\end{aligned} \tag{11c}$$

$$\overline{d}_i = d_i - \dot{z}_{idu} \tag{11d}$$

STEP n: Take the time derivative of s_n to have

$$\begin{aligned}
\dot{s}_n &= d_n + b_n u - \dot{z}_{nd} \\
&= d_n + b_n u - \dot{z}_{ndk} - \dot{z}_{ndu}
\end{aligned} \tag{12}$$

where $\dot{z}_{ndk} = \ddot{z}_{(n-1)dk} - \dot{\hat{\overline{d}}}_{n-1} - c_{n-1}(-c_{n-1}s_{n-1} + s_n - \hat{\overline{d}}_{n-1})$ and $\dot{z}_{ndu} = -c_{n-1}\overline{d}_{n-1}$. Let $\overline{d}_n = d_n - \dot{z}_{ndu}$, then (12) becomes

$$\dot{s}_n = \overline{d}_n + b_m \Delta b u - \dot{z}_{ndk} \tag{13}$$

The controller can thus be designed as

$$u = \frac{1}{b_m}(-\hat{\overline{d}}_n + \dot{z}_{ndk} - c_n s_n - u_r) \tag{14}$$

where $\hat{\overline{d}}_n$ is an estimate of $\overline{d}_n$ and u_r is a robust term to be designed. Substituting (14) into (13) gives

$$\dot{s}_n = \overline{d}_n + \Delta b(-\hat{\overline{d}}_n + \dot{z}_{ndk} - c_n s_n - u_r) - \dot{z}_{ndk}$$

$$= -c_n s_n + (\overline{d}_n - \hat{\overline{d}}_n) + (1 - \Delta b)(\hat{\overline{d}}_n - \dot{z}_{ndk} + c_n s_n) - \Delta b u_r \qquad (15)$$

Let us collect all the error dynamics for subsystems from (5), (11) and (15) to form the error dynamics for the whole closed loop system as

$$\dot{s}_1 = -c_1 s_1 + s_2 + (\overline{d}_1 - \hat{\overline{d}}_1) \qquad (16a)$$

$$\dot{s}_i = -c_i s_i + s_{i+1} + (\overline{d}_i - \hat{\overline{d}}_i), \quad i = 2,...,n-2 \qquad (16b)$$

$$\dot{s}_{n-1} = -c_{n-1} s_{n-1} + s_n + (\overline{d}_{n-1} - \hat{\overline{d}}_{n-1}) \qquad (16c)$$

$$\dot{s}_n = -c_n s_n + (\overline{d}_n - \hat{\overline{d}}_n) + (1 - \Delta b)(\hat{\overline{d}}_n - \dot{z}_{ndk} + c_n s_n) - \Delta b u_r \qquad (16d)$$

where $\overline{d}_1 = d_1$ and $\hat{\overline{d}}_1 = \hat{d}_1$ for notation consistence. Let us look at equation (16d) first. If the robust term u_r can be designed to cover the term $(1 - \Delta b)(\hat{\overline{d}}_n - \dot{z}_{ndk} + c_n s_n)$, and if a proper update law for $\hat{\overline{d}}_n$ is found so that $\hat{\overline{d}}_n \to \overline{d}_n$, then (16d) implies convergence of s_n. If s_n in (16c) is small due to its convergence and an update law for $\hat{\overline{d}}_{n-1}$ is designed so that $\hat{\overline{d}}_{n-1} \to \overline{d}_{n-1}$, then (16c) implies convergence of s_{n-1}. Likewise, we may have convergence of s_1 in (16a) if s_2 is small and an update law for $\hat{\overline{d}}_1$ is found to have $\hat{\overline{d}}_1 \to \overline{d}_1$. Therefore, we need to design a robust term u_r and update laws for $\hat{\overline{d}}_i$, $i = 1,...,n$ to ensure convergence of all states. It can be observed that the error dynamics for each subsystem contains an error term between the lumped uncertainty and its estimate. Since these uncertainties enter the system in a mismatched fashion, so they are mismatched uncertainties. Besides, these uncertainties are time-varying and their variation bounds are not available. These restrictions invalidate most of the robust or adaptive strategies. Here, we would like to employ the function approximation techniques to represent these uncertainties and their estimates as finite combinations of known basis functions as

$$\overline{d}_i = \mathbf{w}_i^T \boldsymbol{\xi}_i + \varepsilon_i, \quad i = 1,\ldots,n \tag{17a}$$

$$\hat{\overline{d}}_i = \hat{\mathbf{w}}_i^T \boldsymbol{\xi}_i, \quad i = 1,\ldots,n \tag{17b}$$

where $\boldsymbol{\xi}_i$ is a vector of basis functions, $\mathbf{w}_i$ and $\hat{\mathbf{w}}_i$ are respectively the coefficient vector and its estimate, and ε_i is the approximation error.

Error surface n: With the representation (17), we may rewrite (16d) as

$$\dot{s}_n = -c_n s_n + \tilde{\mathbf{w}}_n^T \boldsymbol{\xi}_n + \varepsilon_n + (1 - \Delta b)(\hat{\overline{d}}_n - \dot{z}_{ndk} + c_n s_n) - \Delta b u_r \tag{18}$$

where $\tilde{\mathbf{w}}_n = \mathbf{w}_n - \hat{\mathbf{w}}_n$ is the approximation error of the coefficient vector. Define a Lyapunov function candidate for the nth error surface

$$V_n = \frac{1}{2} s_n^2 + \frac{1}{2} \tilde{\mathbf{w}}_n^T \boldsymbol{\Gamma}_n \tilde{\mathbf{w}}_n \tag{19}$$

where $\boldsymbol{\Gamma}_n$ is a positive definite matrix. Taking the time derivative of (19) along the trajectory of (18), we have

$$\dot{V}_n = s_n \dot{s}_n - \tilde{\mathbf{w}}_n^T \boldsymbol{\Gamma}_n \dot{\hat{\mathbf{w}}}_n$$

$$= s_n[-c_n s_n + \tilde{\mathbf{w}}_n^T \boldsymbol{\xi}_n + \varepsilon_n + (1 - \Delta b)(\hat{\overline{d}}_n - \dot{z}_{ndk} + c_n s_n) - \Delta b u_r] - \tilde{\mathbf{w}}_n^T \boldsymbol{\Gamma}_n \dot{\hat{\mathbf{w}}}_n$$

$$= -c_n s_n^2 + s_n[\varepsilon_n + (1 - \Delta b)(\hat{\overline{d}}_n - \dot{z}_{ndk} + c_n s_n) - \Delta b u_r] + \tilde{\mathbf{w}}_n^T (s_n \boldsymbol{\xi}_n - \boldsymbol{\Gamma}_n \dot{\hat{\mathbf{w}}}_n)$$

$$\leq -c_n s_n^2 + s_n \varepsilon_n + (1 + \beta_{max}) \left| \hat{\overline{d}}_n - \dot{z}_{ndk} + c_n s_n \right| |s_n| - \Delta b s_n u_r + \tilde{\mathbf{w}}_n^T (s_n \boldsymbol{\xi}_n - \boldsymbol{\Gamma}_n \dot{\hat{\mathbf{w}}}_n)$$

Pick the robust term and the update law as

$$u_r = \frac{1 + \beta_{max}}{\beta_{min}} \left| \hat{\overline{d}}_n - \dot{z}_{ndk} + c_n s_n \right| \mathrm{sgn}(s_n) \tag{20a}$$

$$\dot{\hat{\mathbf{w}}}_n = \boldsymbol{\Gamma}_n^{-1}(s_n \boldsymbol{\xi}_n - \sigma_n \hat{\mathbf{w}}_n) \tag{20b}$$

where the term with $\sigma_n > 0$ is a σ-modification term to robustify the update law. Then we have

$$\dot{V}_n \le -c_n s_n^2 + s_n \varepsilon_n + \sigma_n \tilde{\mathbf{w}}_n^T \hat{\mathbf{w}}_n$$
$$= -c_n s_n^2 + s_n \varepsilon_n + \sigma_n \tilde{\mathbf{w}}_n^T (\mathbf{w}_n - \tilde{\mathbf{w}}_n) \tag{21}$$
$$\le \underbrace{-c_n s_n^2 + |s_n||\varepsilon_n|}_{(a)} + \sigma_n \underbrace{(\tilde{\mathbf{w}}_n^T \mathbf{w}_n - \|\tilde{\mathbf{w}}_n\|^2)}_{(b)}$$

Part (a) in (21) is further derived as

$$-c_n s_n^2 + |s_n||\varepsilon_n| = -\frac{1}{2}\left(\sqrt{c_n}|s_n| - \frac{|\varepsilon_n|}{\sqrt{c_n}}\right)^2 - \frac{1}{2}\left(c_n s_n^2 - \frac{\varepsilon_n^2}{c_n}\right)$$
$$\le -\frac{1}{2}\left(c_n s_n^2 - \frac{\varepsilon_n^2}{c_n}\right)$$

Part (b) is then derived to be

$$\tilde{\mathbf{w}}_n^T \mathbf{w}_n - \|\tilde{\mathbf{w}}_n\|^2 \le \|\tilde{\mathbf{w}}_n\|\|\mathbf{w}_n\| - \|\tilde{\mathbf{w}}_n\|^2$$
$$= -\frac{1}{2}(\|\tilde{\mathbf{w}}_n\| - \|\mathbf{w}_n\|)^2 - \frac{1}{2}(\|\tilde{\mathbf{w}}_n\|^2 - \|\mathbf{w}_n\|^2)$$
$$\le -\frac{1}{2}(\|\tilde{\mathbf{w}}_n\|^2 - \|\mathbf{w}_n\|^2)$$

Therefore, (19) becomes

$$\dot{V}_n \le -\frac{1}{2}\left(c_n s_n^2 - \frac{\varepsilon_n^2}{c_n}\right) - \frac{\sigma_n}{2}(\|\tilde{\mathbf{w}}_n\|^2 - \|\mathbf{w}_n\|^2)$$
$$= \underbrace{-\frac{c_n}{2}s_n^2 - \frac{\sigma_n}{2}\|\tilde{\mathbf{w}}_n\|^2}_{(c)} + \frac{\sigma_n}{2}\|\mathbf{w}_n\|^2 + \frac{1}{2c_n}\varepsilon_n^2 \tag{22}$$

We would like to relate part (c) in (22) to V_n by considering (19) as

$$V_n = \frac{1}{2}s_n^2 + \frac{1}{2}\tilde{\mathbf{w}}_n^T \Gamma_n \tilde{\mathbf{w}}_n \le \frac{1}{2}s_n^2 + \frac{1}{2}\lambda_{max}(\Gamma_n)\|\tilde{\mathbf{w}}_n\|^2$$

Now, (22) can be further derived as

$$\dot{V}_n \le \pm \alpha_n V_n - \left[\frac{c_n}{2} s_n^2 + \frac{\sigma_n}{2} \left\| \tilde{\mathbf{w}}_n \right\|^2 \right] + \frac{\sigma_n}{2} \left\| \mathbf{w}_n \right\|^2 + \frac{1}{2c_n} \varepsilon_n^2$$

$$\le -\alpha_n V_n + \alpha_n \left[\frac{1}{2} s_n^2 + \frac{1}{2} \lambda_{max}(\mathbf{\Gamma}_n) \left\| \tilde{\mathbf{w}}_n \right\|^2 \right] - \left[\frac{c_n}{2} s_n^2 + \frac{\sigma_n}{2} \left\| \tilde{\mathbf{w}}_n \right\|^2 \right]$$

$$+ \frac{\sigma_n}{2} \left\| \mathbf{w}_n \right\|^2 + \frac{1}{2c_n} \varepsilon_n^2$$

$$= -\alpha_n V_n + \left[\frac{\alpha_n}{2} - \frac{c_n}{2} \right] s_n^2 + \left[\frac{\alpha_n}{2} \lambda_{max}(\mathbf{\Gamma}_n) - \frac{\sigma_n}{2} \right] \left\| \tilde{\mathbf{w}}_n \right\|^2$$

$$+ \frac{\sigma_n}{2} \left\| \mathbf{w}_n \right\|^2 + \frac{1}{2c_n} \varepsilon_n^2$$

Pick $\alpha_n \le \min \left\{ c_n, \dfrac{\sigma_n}{\lambda_{max}(\mathbf{\Gamma}_n)} \right\}$, then we have

$$\dot{V}_n \le -\alpha_n V_n + \frac{\sigma_n}{2} \left\| \mathbf{w}_n \right\|^2 + \frac{1}{2c_n} \varepsilon_n^2 \tag{23}$$

Hence, $\dot{V}_n < 0$ whenever

$$(s_n, \tilde{\mathbf{w}}_n) \in E_n \equiv \left\{ (s_n, \tilde{\mathbf{w}}_n) \middle| V_n > \phi_n \equiv \frac{\sigma_n}{2\alpha_n} \left\| \mathbf{w}_n \right\|^2 + \frac{1}{2c_n \alpha_n} \sup_{\tau \ge t_0} \varepsilon_n^2(\tau) \right\} \tag{24}$$

where ϕ_n can be regarded as the thickness of the boundary layer of the nth surface. Therefore, when outside the boundary layer, we may have $\dot{V}_n < 0$ and s_n will converge to the boundary layer. The result obtained here implies that $(s_n, \tilde{\mathbf{w}}_n)$ is uniformly ultimately bounded. In addition, we can conclude that given any $\mu_n > 0$ there exist $T_n \ge t_0 \ge 0$ such that

$$V_n(t) \le \phi_n + \mu_n \quad \text{for} \quad t \ge T_n \tag{25}$$

Error surface n-1: By using the function approximation, (16c) can be represented as

$$\dot{s}_{n-1} = -c_{n-1} s_{n-1} + s_n + \tilde{\mathbf{w}}_{n-1}^T \boldsymbol{\xi}_{n-1} + \varepsilon_{n-1} \tag{26}$$

Define the Lyapunov function candidate

$$V_{n-1} = \frac{1}{2}s_{n-1}^2 + \frac{1}{2}\tilde{\mathbf{w}}_{n-1}^T \mathbf{\Gamma}_{n-1}\tilde{\mathbf{w}}_{n-1} \tag{27}$$

where $\mathbf{\Gamma}_{n-1}$ is a positive definite matrix. Taking the time derivative of (27) along the trajectory (26), we have

$$
\begin{aligned}
\dot{V}_{n-1} &= s_{n-1}\dot{s}_{n-1} - \tilde{\mathbf{w}}_{n-1}^T \mathbf{\Gamma}_{n-1}\dot{\hat{\mathbf{w}}}_{n-1} \\
&= s_{n-1}(-c_{n-1}s_{n-1} + s_n + \tilde{\mathbf{w}}_{n-1}^T \boldsymbol{\xi}_{n-1} + \varepsilon_{n-1}) - \tilde{\mathbf{w}}_{n-1}^T \mathbf{\Gamma}_{n-1}\dot{\hat{\mathbf{w}}}_{n-1} \\
&= -c_{n-1}s_{n-1}^2 + s_{n-1}(s_n + \varepsilon_{n-1}) + \tilde{\mathbf{w}}_{n-1}^T(s_{n-1}\boldsymbol{\xi}_{n-1} - \mathbf{\Gamma}_{n-1}\dot{\hat{\mathbf{w}}}_{n-1})
\end{aligned}
$$

Select the update law as

$$\dot{\hat{\mathbf{w}}}_{n-1} = \mathbf{\Gamma}_{n-1}^{-1}(s_{n-1}\boldsymbol{\xi}_{n-1} - \sigma_{n-1}\hat{\mathbf{w}}_{n-1}) \tag{28}$$

where $\sigma_{n-1} > 0$. Then we may further have

$$
\begin{aligned}
\dot{V}_{n-1} &= -c_{n-1}s_{n-1}^2 + s_{n-1}(s_n + \varepsilon_{n-1}) + \sigma_{n-1}\tilde{\mathbf{w}}_{n-1}^T \hat{\mathbf{w}}_{n-1} \\
&= -c_{n-1}s_{n-1}^2 + s_{n-1}(s_n + \varepsilon_{n-1}) + \sigma_{n-1}(\tilde{\mathbf{w}}_{n-1}^T \mathbf{w}_{n-1} - \|\tilde{\mathbf{w}}_{n-1}\|^2) \\
&\leq -\frac{1}{2}\left[c_{n-1}s_{n-1}^2 - \frac{1}{c_{n-1}}(s_n + \varepsilon_{n-1})^2\right] - \frac{\sigma_{n-1}}{2}(\|\tilde{\mathbf{w}}_{n-1}\|^2 - \|\mathbf{w}_{n-1}\|^2) \\
&= \underbrace{-\frac{c_{n-1}}{2}s_{n-1}^2 - \frac{\sigma_{n-1}}{2}\|\tilde{\mathbf{w}}_{n-1}\|^2}_{(d)} + \frac{1}{2c_{n-1}}(s_n + \varepsilon_{n-1})^2 + \frac{\sigma_{n-1}}{2}\|\mathbf{w}_{n-1}\|^2
\end{aligned}
$$

Relate (d) in the above equation to V_{n-1} to have

$$\dot{V}_{n-1} \le -\alpha_{n-1}V_{n-1} + \alpha_{n-1}\left[\frac{1}{2}s_{n-1}^2 + \frac{1}{2}\lambda_{\max}(\boldsymbol{\Gamma}_{n-1})\|\tilde{\mathbf{w}}_{n-1}\|^2\right] - \frac{c_{n-1}}{2}s_{n-1}^2$$

$$-\frac{\sigma_{n-1}}{2}\|\tilde{\mathbf{w}}_{n-1}\|^2 + \frac{\sigma_{n-1}}{2}\|\mathbf{w}_{n-1}\|^2 + \frac{1}{2c_{n-1}}(\varepsilon_{n-1}+s_n)^2$$

$$= -\alpha_{n-1}V_{n-1} + \left[\frac{\alpha_{n-1}}{2} - \frac{c_{n-1}}{2}\right]s_{n-1}^2 + \left[\frac{\alpha_{n-1}}{2}\lambda_{\max}(\boldsymbol{\Gamma}_{n-1}) - \frac{\sigma_{n-1}}{2}\right]\|\tilde{\mathbf{w}}_{n-1}\|^2$$

$$+\frac{\sigma_{n-1}}{2}\|\mathbf{w}_{n-1}\|^2 + \frac{1}{2c_{n-1}}(\varepsilon_{n-1}+s_n)^2$$

Pick $\alpha_{n-1} \le \min\left\{c_{n-1}, \dfrac{\sigma_{n-1}}{\lambda_{\max}(\boldsymbol{\Gamma}_{n-1})}\right\}$, then we further have

$$\dot{V}_{n-1} \le -\alpha_{n-1}V_{n-1} + \frac{\sigma_{n-1}}{2}\|\mathbf{w}_{n-1}\|^2 + \frac{1}{2c_{n-1}}(\varepsilon_{n-1}+s_n)^2 \qquad (29)$$

To determine the definiteness of (29) for $t \ge t_0$ two cases are considered: one is before the convergence of V_n and the other is after convergence of V_n. This way, we may clarify the convergence order of s_{n-1} to s_n, and then we may further ensure the order of convergence of the error signals $s_1, s_2, ..., s_n$ in the error space later.

CASE 1: $t_0 \le t < T_n$ (i.e., before convergence of V_n)

Firstly, we claim that the inequality $s_n^2 \le 2\max\{V_n(t_0), \phi_n + \mu_n\}$ is valid. This is proved by considering equation (19) where we have

$$s_n^2(t) = 2V_n(t) - \tilde{\mathbf{w}}_n^T \boldsymbol{\Gamma}_n \tilde{\mathbf{w}}_n \le 2V_n(t) \le 2V_n(t_0)$$

In addition, (25) gives $\dfrac{1}{2}s_n^2 + \dfrac{1}{2}\tilde{\mathbf{w}}_n^T \boldsymbol{\Gamma}_n \tilde{\mathbf{w}}_n \le \phi_n + \mu_n \ \forall t \ge T_n$. This implies $\dfrac{1}{2}s_n^2 \le \phi_n + \mu_n$ for all $t \ge T_n - \tau_n$ for some $\tau_n \ge 0$, or we may rewrite the inequality as $s_n^2 \le 2(\phi_n + \mu_n) \ \forall t \ge T_n - \tau_n$. This completes the proof of the claim. From (29) and the claim, we may have

$$\dot{V}_{n-1} \leq -\alpha_{n-1}V_{n-1} + \frac{\sigma_{n-1}}{2}\|\mathbf{w}_{n-1}\|^2 + \frac{1}{2c_{n-1}}[|\varepsilon_{n-1}| + \sqrt{2\max\{V_n(t_0), \phi_n + \mu_n\}}]^2 \quad (30)$$

The right hand side of (30) will be less than zero whenever $V_{n-1} > \phi_{n-1}$ where

$$\phi_{n-1} \equiv \frac{\sigma_{n-1}}{2\alpha_{n-1}}\|\mathbf{w}_{n-1}\|^2 + \frac{1}{2c_{n-1}\alpha_{n-1}}[\sup_{\tau \geq t_0}|\varepsilon_{n-1}(\tau)| + \sqrt{2\max\{V_n(t_0), \phi_n + \mu_n\}}]^2 \quad (31)$$

i.e., $\dot{V}_{n-1} < 0$ whenever

$$(s_{n-1}, \tilde{\mathbf{w}}_{n-1}) \in E_{n-1} \equiv \{(s_{n-1}, \tilde{\mathbf{w}}_{n-1})|V_{n-1} > \phi_{n-1}\} \quad (32)$$

Therefore, we have proved that V_{n-1} is bounded for all $t \in [t_0, T_n]$, i.e., before convergence of V_n, V_{n-1} is bounded.

CASE 2: $t \geq T_n$ (i.e., after convergence of V_n)

In this case, V_n has converged into $\phi_n + \mu_n$, while V_{n-1} is bounded. Therefore, $(s_{n-1}, \tilde{\mathbf{w}}_{n-1})$ is uniformly ultimately bounded. Furthermore, define

$$\phi'_{n-1} \equiv \frac{\sigma_{n-1}}{2\alpha_{n-1}}\|\mathbf{w}_{n-1}\|^2 + \frac{1}{2c_{n-1}\alpha_{n-1}}[\sup_{\tau \geq t_0}|\varepsilon_{n-1}(\tau)| + \sqrt{2(\phi_n + \mu_n)}]^2$$

then it can be proved that given any $\mu_{n-1} > 0$, there exist $T_{n-1} \geq T_n$ such that

$$V_{n-1} \leq \phi'_{n-1} + \mu_{n-1} \quad \forall t \geq T_{n-1} \quad (33)$$

From these two cases, we obtain boundedness of $(s_{n-1}, \tilde{\mathbf{w}}_{n-1})$. Let us derive similarly to all other error surfaces below.

Error surface i: $i = n-2, n-3, \ldots, 1$

The error dynamics for the ith surface with function approximation representation to the uncertain terms can be written as

$$\dot{s}_i = -c_i s_i + s_{i+1} + \tilde{\mathbf{w}}_i^T \xi_i + \varepsilon_i \quad (34)$$

The Lyapunov function candidate for the ith error surface is

$$V_i = \frac{1}{2} s_i^2 + \frac{1}{2} \tilde{\mathbf{w}}_i^T \boldsymbol{\Gamma}_i \tilde{\mathbf{w}}_i$$

(35)

where $\boldsymbol{\Gamma}_i$ is a positive definite matrix. The time derivative of V_i is computed as

$$\dot{V}_i = -c_i s_i^2 + s_i (s_{i+1} + \varepsilon_i) + \tilde{\mathbf{w}}_i^T (s_i \boldsymbol{\xi}_i - \boldsymbol{\Gamma}_i \dot{\hat{\mathbf{w}}}_i)$$

The update law is designed to be

$$\dot{\hat{\mathbf{w}}}_i = \boldsymbol{\Gamma}_i^{-1} (s_i \boldsymbol{\xi}_i - \sigma_i \hat{\mathbf{w}}_i) \quad \sigma_i > 0$$

(36)

After some rearrangements, the time derivative of V_i becomes the form similar to the one in inequality (29)

$$\dot{V}_i \le -\alpha_i V_i + \frac{\sigma_i}{2} \|\mathbf{w}_i\|^2 + \frac{1}{2c_i} (\varepsilon_i + s_{i+1})^2$$

where $\alpha_i \le \min \left\{ c_i, \dfrac{\sigma_i}{\lambda_{\max}(\boldsymbol{\Gamma}_i)} \right\}$. For $t_0 \le t < T_{i+1}$, we may have $\dot{V}_i < 0$ whenever

$$(s_i, \tilde{\mathbf{w}}_i) \in E_i \equiv \{ (s_i, \tilde{\mathbf{w}}_i) | V_i > \phi_i \}$$

(37)

where $\phi_i \equiv \dfrac{\sigma_i}{2\alpha_i} \|\mathbf{w}_i\|^2 + \dfrac{1}{2c_i\alpha_i} [\sup_{\tau \ge t_0} |\varepsilon_i(\tau)| + \sqrt{2 \max\{V_{i+1}(t_0), \phi_{i+1} + \mu_{i+1}\}}]^2$

This implies that V_i is bounded for all $t \in [t_0, T_{i+1}]$, i.e., before convergence of V_{i+1}, V_i is bounded. After the convergence of V_{i+1} (i.e., $t \ge T_{i+1}$) into $\phi'_{i+1} + \mu_{i+1}$, V_i is bounded. We may conclude that $(s_i, \tilde{\mathbf{w}}_i)$ is uniformly ultimately bounded. Furthermore, it can be proved that given any $\mu_i > 0$, there exist $T_i \ge T_{i+1}$ such that $V_i \le \phi'_i + \mu_i$ for all $t \ge T_i$, where

$$\phi_i' \equiv \frac{\sigma_i}{2\alpha_i} \|\mathbf{w}_i\|^2 + \frac{1}{2c_i\alpha_i} [\sup_{\tau \geq t_0} |\varepsilon_i(\tau)| + \sqrt{2(\phi_{i+1} + \mu_{i+1})}]^2 \qquad (38)$$

Therefore, we have established the order of convergence starting from V_n and ending with V_1, and this further implies the convergence order starting form s_n and ending with s_1. During convergence of s_i, boundedness of $s_j, j = i-1,...,1$ are ensured. Specifically, when $t \geq T_1$, we may have

$$|s_1(t)| = |z_1 - z_{1d}| = \sqrt{2V_1 - \tilde{\mathbf{w}}_1^T \mathbf{\Gamma}_1 \tilde{\mathbf{w}}_1} \leq \sqrt{2(\phi_1' + \mu_1) - \lambda_{\min}(\mathbf{\Gamma}_1)\|\tilde{\mathbf{w}}_1\|^2} \qquad (39)$$

The above derivation only demonstrates the boundedness of the closed loop system, but in practical applications the transient performance is also of great importance. This can be analyzed by respectively deriving the bounds of the error signals for each surface. With these error bounds, the transient performance can be further ensured. Let us start from the last error surface followed by other surfaces.

Error surface *n*: Applying the comparison lemma to (23) gives the upper bound for V_n as

$$V_n \leq e^{-\alpha_n(t-t_0)} V_n(t_0) + \frac{\sigma_n}{2\alpha_n} \|\mathbf{w}_n\|^2 [1 - e^{-\alpha_n(t-t_0)}]$$
$$+ \frac{1}{2c_n} \int_{t_0}^{t} e^{-\alpha_n(t-\tau)} \varepsilon_n^2(\tau) d\tau \qquad (40)$$

The integration can be simplified with

$$\int_{t_0}^{t} e^{-\alpha_n(t-\tau)} \varepsilon_n^2(\tau) d\tau \leq \sup_{t_0 \leq \tau \leq t} \varepsilon_n^2(\tau) \int_{t_0}^{t} e^{-\alpha_n(t-\tau)} d\tau$$
$$= \sup_{t_0 \leq \tau \leq t} \varepsilon_n^2(\tau) \frac{1 - e^{-\alpha_n(t-t_0)}}{\alpha_n} \leq \frac{1}{\alpha_n} \sup_{t_0 \leq \tau \leq t} \varepsilon_n^2(\tau)$$

Hence, (40) becomes

$$V_n \le e^{-\alpha_n(t-t_0)}V_n(t_0) + \frac{\sigma_n}{2\alpha_n}\|\mathbf{w}_n\|^2 + \frac{1}{2c_n\alpha_n}\sup_{t_0 \le \tau \le t}\varepsilon_n^2(\tau)$$

$$= e^{-\alpha_n(t-t_0)}V_n(t_0) + \phi_n$$

The bound for s_n^2 can thus be derived as

$$s_n^2(t) = 2V_n(t) - \tilde{\mathbf{w}}_n^T\boldsymbol{\Gamma}_n\tilde{\mathbf{w}}_n$$

$$\le 2[e^{-\alpha_n(t-t_0)}V_n(t_0) + \phi_n] - \lambda_{\min}(\boldsymbol{\Gamma}_n)\|\tilde{\mathbf{w}}_n\|^2$$

$$\le 2e^{-\alpha_n(t-t_0)}V_n(t_0) + 2\phi_n$$

Finally, we have $|s_n(t)| \le \sqrt{2V_n(t_0)} + \sqrt{2\phi_n}$.

Error surface i: $i = n-1, n-2, ..., 1$

It can be easily derived with $V_i(t) \le e^{-\alpha_n(t-t_0)}V_i(t_0) + \phi_i$, and the bounds for the error signals are with $|s_i(t)| \le \sqrt{2V_i(t_0)} + \sqrt{2\phi_i}$.

It is observed that each and every error bound has its own dominant control parameters; therefore, a compromise among tracking accuracies, estimation errors and control efforts may be made by tuning these parameters.

4.3 Summary

In this section, we summarize the results we have from chapter 3 and this chapter. It starts from considering an n-th order underactuated system in the X-space represented in (3.51) as

$$\begin{aligned}
\dot{x}_1 &= x_2 \\
\dot{x}_2 &= f_2(\mathbf{x}) + b_2(\mathbf{x})u \\
\dot{x}_3 &= x_4 \\
\dot{x}_4 &= f_4(\mathbf{x}) + b_4(\mathbf{x})u \\
&\vdots \\
\dot{x}_{n-1} &= x_n \\
\dot{x}_n &= f_n(\mathbf{x}) + b_n(\mathbf{x})u
\end{aligned} \qquad (41)$$

where n is an even number. The functions $f_i(\mathbf{x}), i = 2,4,...,n$ are unknown, while $b_i(\mathbf{x}), i = 2,4,...,n$ are known to ensure the coordinate transformations shown in (3.52) to be feasible.

$$z_1 = x_1 - \int_0^{x_3} \frac{b_2(s)}{b_4(s)} ds$$

$$z_2 = x_2 - \frac{b_2(\mathbf{x})}{b_4(\mathbf{x})} x_4$$

$$z_3 = x_3 - \int_0^{x_5} \frac{b_4(s)}{b_6(s)} ds$$

$$z_4 = x_4 - \frac{b_4(\mathbf{x})}{b_6(\mathbf{x})} x_4$$

$$\vdots \tag{42}$$

$$z_{n-3} = x_{n-3} - \int_0^{x_{n-1}} \frac{b_{n-2}(s)}{b_n(s)} ds$$

$$z_{n-2} = x_{n-2} - \frac{b_{n-2}(\mathbf{x})}{b_n(\mathbf{x})} x_n$$

$$z_{n-1} = x_{n-1}$$

$$z_n = x_n$$

By using this coordinate transformation, we may represent the system in the Z-space in (3.55) as

$$\dot{z}_1 = z_2 + d_1(\mathbf{z})$$

$$\dot{z}_2 = z_3 + d_2(\mathbf{z})$$

$$\vdots$$

$$\dot{z}_{n-3} = z_{n-2} + d_{n-3}(\mathbf{z}) \tag{43}$$

$$\dot{z}_{n-2} = z_{n-1} + d_{n-2}(\mathbf{z})$$

$$\dot{z}_{n-1} = z_n + d_{n-1}(\mathbf{z})$$

$$\dot{z}_n = d_n(\mathbf{z}) + b_n(\mathbf{z})u$$

where we assume the functions $d_i(\mathbf{z}), i = 1, 2, ..., n$ to be unavailable. Although we know $b_n(\mathbf{z})$, the following control strategy is derived to allow it to be unknown.

A set of error signals are defined in (2) as

$$
\begin{aligned}
s_1 &= z_1 - z_{1d} \\
s_2 &= z_2 - z_{2d} \\
&\;\vdots \\
s_n &= z_n - z_{nd}
\end{aligned}
\tag{44}
$$

By selecting the desired trajectories (4) and (11)

$$z_{2d} = \dot{z}_{1d} - \hat{d}_1 - c_1 s_1 \tag{45a}$$

$$z_{(i+1)d} = \dot{z}_{idk} - \hat{d}_i - c_i s_i, \quad i = 2, ..., n-1 \tag{45b}$$

and the controller in (14)

$$u = \frac{1}{b_m}(-\hat{d}_n + \dot{z}_{ndk} - c_n s_n - u_r) \tag{46}$$

the error dynamics becomes the one in (16)

$$\dot{s}_1 = -c_1 s_1 + s_2 + (d_1 - \hat{d}_1) \tag{47a}$$

$$\dot{s}_i = -c_i s_i + s_{i+1} + (\bar{d}_i - \hat{d}_i), \quad i = 2, ..., n-2 \tag{47b}$$

$$\dot{s}_{n-1} = -c_{n-1} s_{n-1} + s_n + (\bar{d}_{n-1} - \hat{d}_{n-1}) \tag{47c}$$

$$\dot{s}_n = -c_n s_n + (\bar{d}_n - \hat{d}_n) + (1 - \Delta b)(\hat{d}_n - \dot{z}_{ndk} + c_n s_n) - \Delta b u_r \tag{47d}$$

where the robust term is selected in (20a)

$$u_r = \frac{1 + \beta_{max}}{\beta_{min}} \left| \hat{d}_n - \dot{z}_{ndk} + c_n s_n \right| \operatorname{sgn}(s_n) \tag{48}$$

The uncertainties are approximated in (17) as

$$\overline{d}_i = \mathbf{w}_i^T \boldsymbol{\xi}_i + \varepsilon_i, \quad i = 1,...,n \qquad (49\text{a})$$

$$\hat{\overline{d}}_i = \hat{\mathbf{w}}_i^T \boldsymbol{\xi}_i, \quad i = 1,...,n \qquad (49\text{b})$$

Their update laws are constructed in (20b), (28) and (36)

$$\dot{\hat{\mathbf{w}}}_i = \boldsymbol{\Gamma}_i^{-1}(s_i \boldsymbol{\xi}_i - \sigma_i \hat{\mathbf{w}}_i) \quad \sigma_i > 0 \qquad (50)$$

where $i = 1,...,n$.

Since many benchmark underactuated systems can be represented in the form of (41), the controller designed here is able to stabilize these systems. In the following chapters, we shall use the same control strategy summarized here to control various uncertain underactuated mechanical systems.

Chapter 5

Cart Pole System

A cart pole system consists of an inverted pendulum mounted on a motor driven cart as shown in Fig. 5.1. It is underactuated because both the cart position x and the pendulum angle θ are controlled by the control u on the cart. This system has been widely studied with various control strategies. Many undergraduate control textbooks, e.g., Ogata (1997), use this system as a benchmark problem. They linearize the system dynamics around the vertical position of the pendulum firstly, and then verify controllability of the system based on the linearized model. A state feedback controller is designed to stabilize the system under the assumption that the angle of the pendulum is small to ensure validity of the linearization. To have a wider range of operation, a nonlinear control scheme is needed. If the system contains uncertainties, some sophisticated controllers have to be designed to give desired performance.

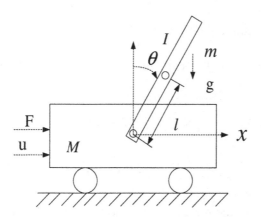

Figure 5.1. A cart pole system

The control problem for the cart pole system is interesting only when the underactuated dynamics is considered and a sufficiently large range of operation is required. To give more challenge to the problem, some system parameters can be assumed to be unavailable and some external disturbances can be included. There are three main control objectives of the cart pole system: (1) stabilization of the pendulum to the vertical position, (2) swing up of the pendulum, (3) tracking problem.

The stabilization of the pendulum starts with the initial angle $\theta(0) \in (-\frac{\pi}{2}, \frac{\pi}{2})$ and the desired angle is at $\theta = 0$. For $|\theta(0)| = \frac{\pi}{2}$ and $\dot\theta(0) \geq 0$, there will be no vertical component of the actuation force to drive the pendulum resulting in the increase of the angle and the system cannot be stabilized. The swing up of the pendulum considers the initial condition $\theta(0) = \pi$ and $\dot\theta(0) = 0$. Mostly, the control strategy includes two stages. The first stage swings up the pendulum to enter the set $(-\frac{\pi}{2}, \frac{\pi}{2})$ followed by a switch to enter the second stage where a regulation controller is implemented. The tracking problem is most difficult among the three problems due to the coupled dynamics between the cart and pendulum. In section 5.1, the dynamic equation of the cart pole system is derived by using the Lagrange equation. The coordinate transformation introduced in chapter 3 is then performed in section 5.2 to represent the system in the Z-space. The last section gives the simulation cases to verify the effectiveness of the design.

5.1 System Dynamics

The kinetic energy of the cart and the pendulum can respectively be represented to be

$$K_{cart} = \frac{1}{2}(M + m)\dot x^2$$

$$K_{pendulum} = \frac{1}{2}(I + ml^2)\dot\theta^2 + ml\dot x\dot\theta\cos\theta$$

where M and m are the masses of the cart and pendulum respectively, l is the distance of the center of gravity of the pendulum to the hinge, and I is the moment of inertial of the pendulum. The total kinetic energy is the combination of K_{cart} and $K_{pendulum}$ as

$$K = K_{cart} + K_{pendulum}$$
$$= \frac{1}{2}(M+m)\dot{x}^2 + \frac{1}{2}(I+ml^2)\dot{\theta}^2 + ml\dot{x}\dot{\theta}\cos\theta \tag{1}$$

With the definition of the potential energy of the pendulum

$$V = mgl(\cos\theta - 1) \tag{2}$$

the Lagrangian can then be found as

$$L = K - V$$
$$= \frac{1}{2}(M+m)\dot{x}^2 + ml\dot{x}\dot{\theta}\cos\theta + \frac{1}{2}(I+ml^2)\dot{\theta}^2 - mgl(\cos\theta - 1) \tag{3}$$

The generalized coordinate vector and the generalized force vector can respectively be assigned as

$$\mathbf{q} = \begin{bmatrix} q_1 \\ q_2 \end{bmatrix} = \begin{bmatrix} x \\ \theta \end{bmatrix} \in \Re^2$$

$$\boldsymbol{\tau} = \begin{bmatrix} F+u \\ 0 \end{bmatrix} \in \Re^2$$

where F is an external disturbance on the cart. Let us compute the quantities below to facilitate the derivation of the equation of motion via the Lagrange equation

$$\frac{\partial L}{\partial \dot{x}} = (M+m)\dot{x} + ml\dot{\theta}\cos\theta$$

$$\frac{\partial L}{\partial x} = 0$$

$$\frac{\partial L}{\partial \dot{\theta}} = ml\dot{x}\cos\theta + (I+ml^2)\dot{\theta} \tag{4}$$

$$\frac{\partial L}{\partial \theta} = mgl\sin\theta - ml\dot{x}\dot{\theta}\sin\theta$$

then we may have the dynamic equation of the cart pole system in the X-space using the Lagrange equation

$$\frac{d}{dt}\frac{\partial L}{\partial \dot{\mathbf{q}}} - \frac{\partial L}{\partial \mathbf{q}} = \tau \tag{5}$$

to have

$$(M + m\sin^2 \theta)\ddot{x} - m\sin\theta(l\dot{\theta}^2 - g\cos\theta) = F + u$$
$$(M + m\sin^2 \theta)l\ddot{\theta} + ml\dot{\theta}^2 \sin\theta\cos\theta - (M + m)g\sin\theta = -\cos\theta(F + u) \tag{6}$$

Define the state vector

$$\mathbf{x} = [x \quad \dot{x} \quad \theta \quad \dot{\theta}]^T = [x_1 \quad x_2 \quad x_3 \quad x_4]^T \in \mathfrak{R}^4$$

and the state space representation becomes

$$\begin{aligned}
\dot{x}_1 &= x_2 \\
\dot{x}_2 &= f_2(\mathbf{x}) + b_2(\mathbf{x})u \\
\dot{x}_3 &= x_4 \\
\dot{x}_4 &= f_4(\mathbf{x}) + b_4(\mathbf{x})u
\end{aligned} \tag{7}$$

where

$$f_2(\mathbf{x}) = \frac{mlx_4^2 \sin x_3 - mg \sin x_3 \cos x_3 + F}{M + m\sin^2 x_3} \tag{8a}$$

$$b_2(x_3) = \frac{1}{M + m\sin^2 x_3} \tag{8b}$$

$$f_4(\mathbf{x}) = \frac{-mlx_4^2 \sin x_3 \cos x_3 + (M + m)g \sin x_3 - F \cos x_3}{Ml + ml \sin^2 x_3} \tag{8c}$$

$$b_4(x_3) = \frac{-\cos x_3}{Ml + ml \sin^2 x_3} \tag{8d}$$

There is only one control u appears in both the dynamic equations of the cart and the pole in (7) and hence the whole system is underactuated.

5.2 Coordinate Transformation

It is apparently that (7) is in the standard form (4.41) with $n = 4$. We may then transform the system from the X-space into Z-space by using the mapping (4.42) as

$$
z_1 = x_1 - \int_0^{x_3} \frac{b_2(s)}{b_4(s)} ds = x_1 - \int_0^{x_3} \frac{\dfrac{1}{M + m \sin^2 s}}{\dfrac{-\cos s}{Ml + ml \sin^2 s}} ds = x_1 + l \ln\left|\sec x_3 + \tan x_3\right|
$$

$$
z_2 = x_2 - \frac{b_2(\mathbf{x})}{b_4(\mathbf{x})} x_4 = x_2 - \frac{\dfrac{1}{M + m \sin^2 x_3}}{\dfrac{-\cos x_3}{Ml + ml \sin^2 x_3}} = x_2 + l x_4 \sec x_3 \tag{9}
$$

$$
z_3 = x_3
$$

$$
z_4 = x_4
$$

It is seen in (9) that the only system parameter appears in the coordinate transformation is the length l from the center of mass of the pendulum to the hinge which can easily be measured practically. All other system parameters can then be allowed to be unknown in both the X-space and Z-space. With the transformation (9), the system dynamics in the Z-space becomes

$$
\begin{aligned}
\dot{z}_1 &= z_2 + d_1(\mathbf{z}) \\
\dot{z}_2 &= z_3 + d_2(\mathbf{z}) \\
\dot{z}_3 &= z_4 + d_3(\mathbf{z}) \\
\dot{z}_4 &= d_4(\mathbf{z}) + b_4(\mathbf{z})u
\end{aligned} \tag{10}
$$

where $d_1(\mathbf{z}) = d_3(\mathbf{z}) = 0$, and

$$
d_2(\mathbf{z}) = -z_3 - l z_4^2 \sec z_3 \tan z_3 + \frac{(M + m)g \tan z_3 - mg \sin z_3 \cos z_3}{M + m \sin^2 z_3} \tag{11a}
$$

$$
d_4(\mathbf{z}) = f_4(\mathbf{z}) = \frac{-ml z_4^2 \sin z_3 \cos z_3 + (M + m)g \sin z_3 - F \cos z_3}{Ml + ml \sin^2 z_3} \tag{11b}
$$

$$b_4(\mathbf{z}) = \frac{-\cos z_3}{Ml + ml\sin^2 z_3} \tag{11c}$$

Since (10) is a special case of (4-41), the controller derived there can be used to stabilize the system regardless of the internal uncertainties and external disturbances. Some simulation cases are presented in the next section to verify the effectiveness of the control strategy.

5.3 Simulation Cases

Let us assume that initially the cart is at 0 (m) and the pendulum is at 60 degrees from rest and the desired position for the cart is at 1 (m) and for the pendulum is 0 degree (vertical position). So, the initial state is $\mathbf{x}(0) = [0 \quad 0 \quad \frac{\pi}{3} \quad 0]^T$ and the desired state is $\mathbf{x}_d = [1 \quad 0 \quad 0 \quad 0]^T$. The actual values of the system parameters in the simulation are $m = 1$ (kg), $M = 2$ (kg), and $l = 0.5$ (m). The controller in (4-46) together with the desired trajectories (4-45) and the robust term (4-48) are used with the parameters selected as $c_1 = 25$, $c_2 = 15$, $c_3 = 5$, and $c_4 = 0.1$. The first 21 terms of the Fourier series are used as the basis functions. The weighting matrices in the update law (4-50) are picked as $\mathbf{\Gamma}_i = 10\mathbf{I}_{21}$, $i = 1,...,4$. The constant σ in (4-50) is set to zero to turn off the σ-modification in the update law. The initial weighting in (4-50) are set to be $\hat{\mathbf{w}}_2(0) = [0.5 \quad 0 \quad \cdots \quad 0]^T \in \mathfrak{R}^{21}$ and $\hat{\mathbf{w}}_4(0) = [0.2 \quad 0 \quad \cdots \quad 0]^T \in \mathfrak{R}^{21}$. Two simulation cases are presented here. The first case is the control of an uncertain inverted pendulum without external disturbances. The second case presents the control result of an uncertain inverted pendulum with the presence of some sinusoidal disturbances on the cart and significant variation in system masses.

CASE 1: Uncertain inverted pendulum
        Supposes the pendulum angle is to be restricted as $z_3 \in [-\frac{\pi}{3}, \frac{\pi}{3}]$, then we may compute the parameters $b_{\min} = 0.3636$, $b_{\max} = 1$, $b_m = 0.603$, $\beta_{\max} = 1.6584$ and $\beta_{\min} = 0.603$. For more conservative consideration, the bounds for z_3 can be selected larger. The simulation results are shown in Figs. 5.2 to 5.5. Fig. 5.2 verifies that both the cart displacement and pendulum angle converge nicely regardless of the uncertainties in the

system dynamics. Fig. 5.3 is the control effort required. Fig. 5.4 presents the convergence of each error surface. Fig. 5.5 shows the function approximation performance. It is seen that all parameters are bounded as desired.

The simulation results show that the decoupling mapping of the underactuated dynamics in the Z-space and the adaptive controller designed in previous chapters are feasible to bring the system to the desired state regardless of the internal uncertainties.

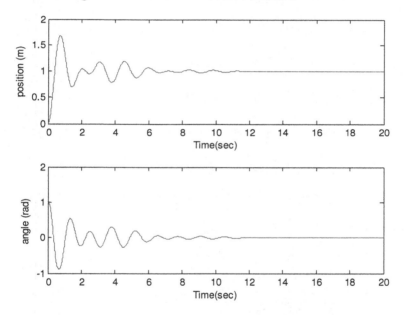

Figure 5.2. Trajectories of (a) cart position (b) pendulum angle

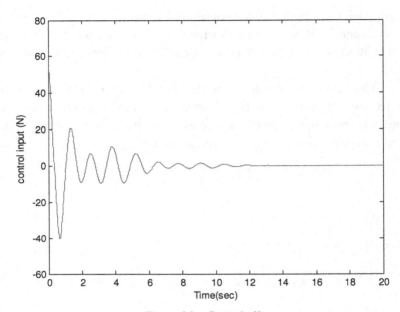

Figure 5.3.   Control effort

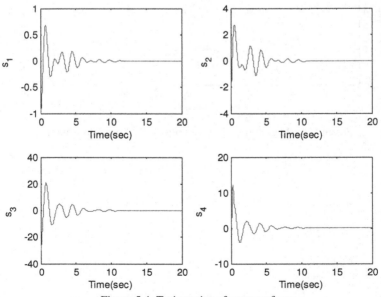

Figure 5.4. Trajectories of error surfaces

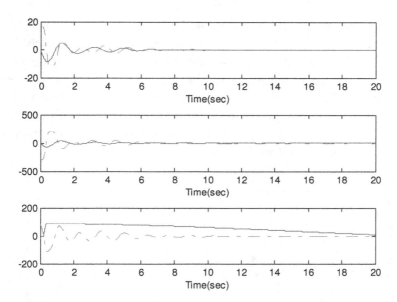

Figure 5.5. Function approximation performance (a) $\overline{d}_2$ (b) $\overline{d}_3$ (c) $\overline{d}_4$

CASE 2: Uncertain inverted pendulum with time-varying masses and disturbances

In this case, in addition to the system uncertainties, the masses of the cart and pendulum are allowed to vary with significant amount as

$$m = 1 + 0.5\sin 3t \ (\text{kg})$$

$$M = 2 + \sin 0.5t \ (\text{kg})$$

A periodic disturbance $F = \cos 2t$ is applied to the cart during $t \in [20,30]$ (sec) to test robustness of the strategy. The controller parameters can be computed to be $b_{\min} = 0.2424$, $b_{\max} = 2$, $b_m = 0.6963$, $\beta_{\max} = 2.8723$ and $\beta_{\min} = 0.3481$. The simulation results are shown in Figs. 5.6 to 5.9. Fig. 5.6 shows that even with time-varying masses and various uncertainties the controller can still bring the system to the desired states. Application of the disturbance to the cart gives significant deviation of the system trajectories during $t \in [20,30]$. However, they are still bounded and go back to the desired states quickly when the external force is removed. Fig. 5.7 is the history of the control effort which shows that even with time-varying masses and external disturbances, the control effort is still with reasonable size. Fig. 5.8 is the convergence of each error surface. Fig. 5.9 shows the function approximation performance.

It is seen that the controller designed in the previous chapter is robust enough to tolerate significant variations in the system parameters as well as the external disturbances.

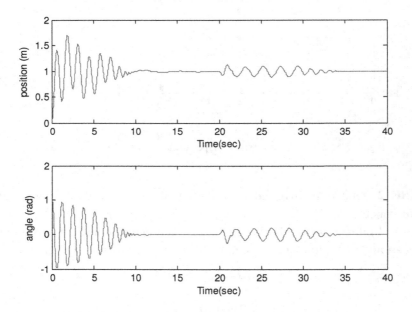

Figure 5.6. Trajectories of (a) cart position (b) pendulum angle

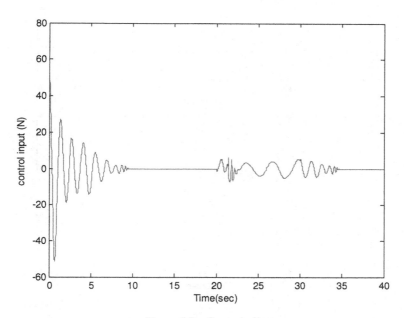

Figure 5.7.   Control effort

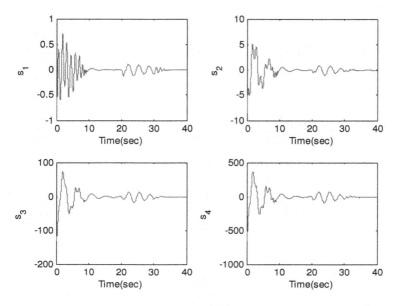

Figure 5.8.   Trajectories of error surfaces

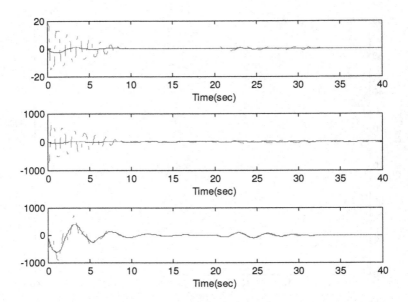

Figure 5.9. Function approximation performance (a) $\overline{d}_2$ (b) $\overline{d}_3$ (c) $\overline{d}_4$

Chapter 6

Overhead Crane

Overhead crane systems (Fig. 6.1) are widely used in the industrial environment to carry heavy loads. The trolley is required to carry a payload connected with a rope to travel a finite distance as fast as possible without overshoot when arriving at the destination. To meet safety requirements, the movement of the payload should be performed with small oscillations. Sometimes the positioning accuracy of the trolley when reaching the destination is also very important. In many situations the hoisting mechanism must lift and lower payloads to avoid obstacles during movement, which may excite oscillations if not properly controlled. In some applications, it is preferable to perform loading/unloading operations during the movement of the crane which will also induce payload swing if not effectively treated. In addition, in some accidents such as in the case when payloads drop down, we would like the controller to be robust enough to ensure stability of the operation. Therefore, consideration of varying payload is a nontrivial attempt for a practical control design in industry.

Most of the existing overhead crane systems are controlled by experienced operators to ensure safety and efficiency. Many controllers have been proposed to control the overhead crane system automatically in recent years (Abdel-Rahman et al., 2003; Ahmad, 2009; Uchiyama, 2009; Yang, 2007). Some of them assume that the system models are precisely available (Balachandran et al., 1999; Collado et al., 2000; Masoud et al., 2004), while some allow the system to contain uncertainties (Bartolini et al., 2002; Lee, 2004; Shyu et al., 2006; Uchiyama, 2009; Yang, 2007). Since the overhead crane system is well-known to be an underactuated system (Chen et al., 2012; Fang, 2003; Shyu, 2006; Oguamanam et al., 2001; Olfati-Saber, 2002; Olfati-Saber and Megretasi, 1998), i.e., one input force at the trolley to control two outputs: the trolley position and the swing angle, the control problem is challenging.

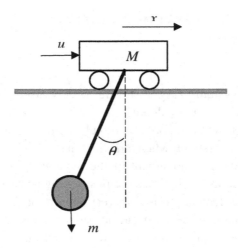

Figure 6.1. An overhead crane system

   In this chapter, we allow the crane to have varying rope length to meet geometric constraints and to allow the payload to have time-varying mass. These two effects are considered as uncertainties in our design, and the control of this crane system becomes extremely difficult. One way to deal with time-varying payload and rope length is to include a robust design which needs the information of the variation bound of the payload (Bartolini et al., 2002; Shyu et al., 2006; Uchiyama, 2009). A conservative strategy is then designed to cover the payload variation which generally consumes a lot of energy. The other possible way to solve the time-varying payload and rope length problem is to apply adaptive design (Yang, 2007; Huang and Chen, 2004b). However, a common assumption for the traditional adaptive scheme is that the uncertainties should be represented in a linearly parameterized form where all unknowns should be collected inside a constant vector. The unknown time-varying payload and rope length cannot always be included as an entry of the unknown constant vector; therefore, the conventional adaptive design is not generally feasible.

Here, we would like to apply the adaptive controller derived in chapter 4 to an overhead crane system after the transformation of the system dynamics into the special cascade form. Section 6.1 derives the dynamic equation of an overhead crane system in the X-space. It is then transformed into the Z-space in section 6.2. Several simulation cases are presented in section 6.3 to justify the controller performance.

6.1 System Dynamics

If we ignore the weight of the rope, then the kinetic energy of the trolley and the payload are found as

$$K_{trolley} = \frac{1}{2}(M+m)\dot{x}^2$$

$$K_{payload} = \frac{1}{2}ml^2\dot{\theta}^2 + ml\dot{x}\dot{\theta}\cos\theta$$

where M is the mass of the trolley, m is the mass of the payload, and l is the length of the rope. The total kinetic energy is then

$$\begin{aligned}
K &= K_{trolley} + K_{payload} \\
&= \frac{1}{2}(M+m)\dot{x}^2 + \frac{1}{2}ml^2\dot{\theta}^2 + ml\dot{x}\dot{\theta}\cos\theta
\end{aligned} \tag{1}$$

The potential energy of the system is found from the height of the payload as

$$V = mgl(\cos\theta - 1) \tag{2}$$

Hence, the Lagrangian is defined to be

$$\begin{aligned}
L &= K - V \\
&= \frac{1}{2}(M+m)\dot{x}^2 + ml\dot{x}\dot{\theta}\cos\theta + \frac{1}{2}ml^2\dot{\theta}^2 - mgl(\cos\theta - 1)
\end{aligned} \tag{3}$$

The generalized coordinate vector and the generalized force vector can respectively be assigned as

$$\mathbf{q} = \begin{bmatrix} q_1 \\ q_2 \end{bmatrix} = \begin{bmatrix} x \\ \theta \end{bmatrix} \in \Re^2$$

$$\tau = \begin{bmatrix} 0 \\ u \end{bmatrix} \in \Re^2$$

Let us compute the following quantities to facilitate the derivation of the equation of motion by using the Lagrange equation

$$\frac{\partial L}{\partial \dot{x}} = (M + m)\dot{x} - ml\dot{\theta}\cos\theta$$

$$\frac{\partial L}{\partial x} = 0$$

$$\frac{\partial L}{\partial \dot{\theta}} = -ml\dot{x}\cos\theta + ml^2\dot{\theta}$$

$$\frac{\partial L}{\partial \theta} = -mgl\sin\theta + ml\dot{x}\dot{\theta}\sin\theta$$

(4)

Then, we may derive the dynamic equation of the overhead crane system by using the Lagrange equation

$$\frac{d}{dt}\frac{\partial L}{\partial \dot{\mathbf{q}}} - \frac{\partial L}{\partial \mathbf{q}} = \tau \tag{5}$$

to have

$$(M + m\sin^2\theta)\ddot{x} + m\sin\theta(l\dot{\theta}^2 + g\cos\theta) = u$$

$$(M + m\sin^2\theta)l\ddot{\theta} - ml\dot{\theta}^2\sin\theta\cos\theta + (M + m)g\sin\theta = -u\cos\theta$$

(6)

Define the state vector

$$\mathbf{x} = [x \quad \dot{x} \quad \theta \quad \dot{\theta}]^T = [x_1 \quad x_2 \quad x_3 \quad x_4]^T \in \Re^4$$

The system (6) can thus be represented in the X-space as

$$\dot{x}_1 = x_2$$
$$\dot{x}_2 = f_2(\mathbf{x}) + b_2(\mathbf{x})u$$
$$\dot{x}_3 = x_4$$
$$\dot{x}_4 = f_4(\mathbf{x}) + b_4(\mathbf{x})u \tag{7}$$

where

$$f_2(\mathbf{x}) = \frac{mlx_4^2 \sin x_3 - mg \sin x_3 \cos x_3}{M + m\sin^2 x_3} \tag{8a}$$

$$b_2(x_3) = \frac{1}{M + m\sin^2 x_3} \tag{8b}$$

$$f_4(\mathbf{x}) = \frac{-mlx_4^2 \sin x_3 \cos x_3 + (M + m)g \sin x_3}{Ml + ml\sin^2 x_3} \tag{8c}$$

$$b_4(x_3) = \frac{-\cos x_3}{Ml + ml\sin^2 x_3} \tag{8d}$$

There is only one control u appears in both the dynamic equations of the trolley and the payload in (7) and hence the whole system is underactuated.

6.2 Coordinate Transformation

It is apparently that (7) is in the standard form (4.41) with $n = 4$. We may then transform the system from X-space into Z-space by using the mapping from (4.42) as

$$z_1 = x_1 - \int_0^{x_3} \frac{b_2(s)}{b_4(s)} ds$$

$$= x_1 - \int_0^{x_3} \frac{\dfrac{M + m\sin^2 s}{-\cos s}}{Ml + ml\sin^2 s} ds = x_1 + l\ln\left|\sec x_3 + \tan x_3\right| \tag{9a}$$

$$z_2 = x_2 - \frac{b_2(\mathbf{x})}{b_4(\mathbf{x})} x_4 = x_2 - \frac{\dfrac{1}{M + m \sin^2 x_3}}{\dfrac{-\cos x_3}{Ml + ml \sin^2 x_3}} = x_2 + lx_4 \sec x_3 \qquad (9b)$$

$$z_3 = x_3 \qquad (9c)$$

$$z_4 = x_4 \qquad (9d)$$

It is seen that in (9) that the only system parameter appears in the coordinate transformation is the rope length l which can easily be measured practically. All other system parameters can then be allowed to be unknown in both the X-space and Z-space. With the transformation (9), the system dynamics in the Z-space becomes

$$\begin{aligned}
\dot{z}_1 &= z_2 + d_1(\mathbf{z}) \\
\dot{z}_2 &= z_3 + d_2(\mathbf{z}) \\
\dot{z}_3 &= z_4 + d_3(\mathbf{z}) \\
\dot{z}_4 &= d_4(\mathbf{z}) + b_4(\mathbf{z})u
\end{aligned} \qquad (10)$$

where $d_1(\mathbf{z}) = d_3(\mathbf{z}) = 0$, and

$$d_2(\mathbf{z}) = -z_3 - lz_4^2 \sec z_3 \tan z_3 + \frac{(M + m)g \tan z_3 - mg \sin z_3 \cos z_3}{M + m \sin^2 z_3}$$

$$d_4(\mathbf{z}) = f_4(\mathbf{z}) = \frac{-mlz_4^2 \sin z_3 \cos z_3 + (M + m)g \sin z_3}{Ml + ml \sin^2 z_3}$$

$$b_4(\mathbf{z}) = \frac{-\cos z_3}{Ml + ml \sin^2 z_3}$$

Since (9) is a special case of (4-41), the controller derived there can be used to stabilize the system regardless of the internal uncertainties and external disturbances. Some simulation cases are presented in the next section to verify the effectiveness of the controller in chapter 4.

6.3 Simulation Cases

Let us consider an overhead crane system in Fig. 6.1 with the nominal parameters $m = 42500$ (kg), $M = 6000$ (kg), and $l = 6$ (m) (Corriga et al., 1998). It is desired to send the trolley from 0(m) to 10(m) in 20 seconds by following a 5-th order profile

$$x_d = y_f \left(\frac{6t^5}{t_f^5} - \frac{15t^4}{t_f^4} + \frac{10t^3}{t_f^3} \right)$$

where $y_f = 10$ (m) and $t_f = 20$ seconds. The reason for selecting a 5-th order profile is to ensure continuity in the acceleration of the motion to have a bumpless transfer from rest to acceleration, deceleration and to rest. To make the problem more challenging, we assume that the payload and the rope length are both time-varying by significant amounts as

$$m = 42500 + 20000\sin t \text{ (kg)}$$

$$l = 6 + 4\cos 0.1t \text{ (m)}$$

The variation of the payload mass is to simulate the loading/unloading operations during the movement, while the changing of the rope length is to allow the obstacle avoidance capability. Three controllers are implemented to compare their performance.

PID Controller:
    A conventional PID controller is designed with the gains $k_p = 0.1$, $k_i = 0.002$, and $k_d = 0.001$. Here, we tried our best to tune the gains for stabilizing the closed loop system. The simulation results are shown in Figs. 6.2 and 6.3. Fig. 6.2(a) is the time history of the cart movement and Fig. 6.2(b) is the rope angle. Since the PID controller is only a second-order linear filter, it is not capable of giving satisfactory performance to the present problem which contains underactuated, highly nonlinear, and time-varying dynamics. Fig. 6.2(a) also shows that the trolley has a one meter magnitude oscillation when arriving at the target position after 10 seconds. This is not an acceptable performance in practical operations, although the control effort is realizable as shown in Fig. 6.3.

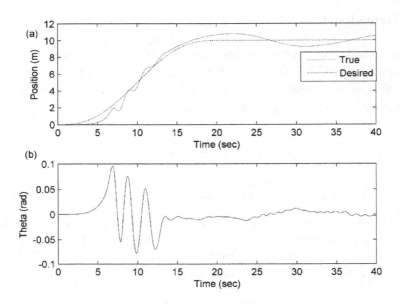

Figure 6.2.   System response of the PID controller. (a) Trolley position (b) Rope angle

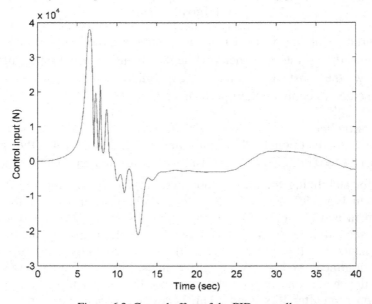

Figure 6.3. Control effort of the PID controller

Proposed Controller:

In this case, we assume that $d_2(\mathbf{z})$, $d_4(\mathbf{z})$ and $b_4(\mathbf{z})$ are uncertainties. Since the maximum rope length $l_{max} = 6 + 4 = 10$ (m) and the maximum payload mass is $m_{max} = 42500 + 20000 = 62500$ (kg), we may have $b_{max} = 8.33 \times 10^{-5}$ and $b_{min} = 1.64 \times 10^{-5}$. Thus we have $b_m = 3.70 \times 10^{-5}$, $\beta_{max} = 2.2509$ and $\beta_{min} = 0.4443$. The controller in (4-46) together with the desired trajectories (4-45) and the robust term (4-48) are used with the parameters $c_1 = 12.733$, $c_2 = 0.7$, $c_3 = 2.08$, and $c_4 = 1.79$. The first 21 terms of the Fourier series are used as the basis functions. The weighting matrices in the update law (4-50) are picked as $\Gamma_i = 10\mathbf{I}_{21}$, $i = 1,...,4$. The constant σ in (4-50) is set to zero to turn off the σ-modification in the update law. The initial weightings in (4-50) are set to be

$$\hat{\mathbf{w}}_2(0) = [1 \quad 0 \quad \cdots \quad 0]^T \in \mathfrak{R}^{21}$$

$$\hat{\mathbf{w}}_3(0) = \hat{\mathbf{w}}_4(0) = [0.005 \quad 0 \quad \cdots \quad 0]^T \in \mathfrak{R}^{21}$$

The simulation results are shown in Figs. 6.4 to 6.7. It is seen in Fig. 6.4(a) that the trolley tracks the desired trajectory nicely regardless of the significant variations of the payload mass and rope length. The rope angle is shown in Fig. 6.4(b) where the maximum deviation is about 1 degree during movement, and a less than 0.3 degree deviation can be seen with convergent tendency after 20 seconds. The control effort of the design is realizable as shown in Fig. 6.5. Therefore, the proposed controller can give good performance in the given problem. Fig. 6.6 shows the trajectories of the error surfaces and Fig. 6.7 is the function approximation performance.

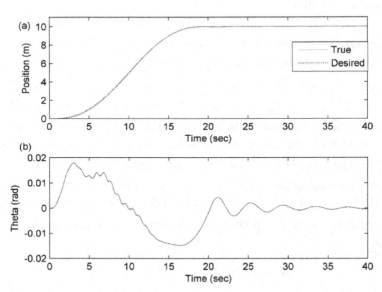

Figure 6.4.   System output with proposed controller. (a) Trolley position (b) Rope angle

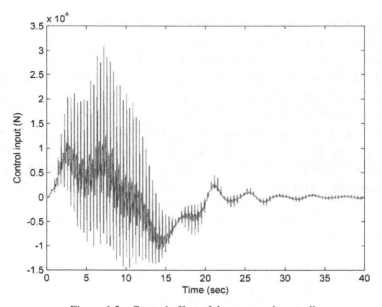

Figure 6.5.   Control effort of the proposed controller

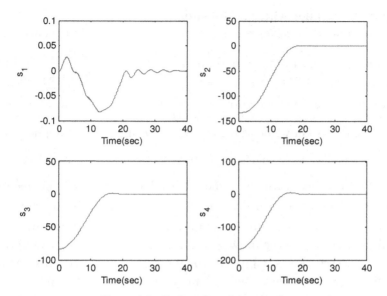

Figure 6.6.   Trajectories of error surfaces

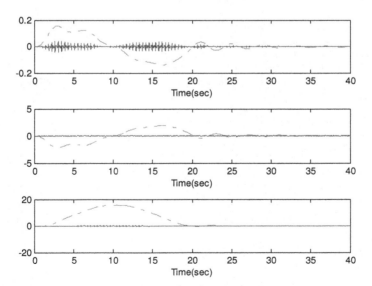

Figure 6.7. Function approximation performance (a) $\overline{d}_2$ (b) $\overline{d}_3$ (c) $\overline{d}_4$

Proposed Controller with Modification:

In our design, a robust term u_r is designed to cover the uncertainties in b_4. From its definition $b_4(z_3) = \dfrac{-\cos z_3}{Ml + ml\sin^2 z_3}$ in (10), we may have the approximation $b_4 \approx \dfrac{-1}{Ml}$ when z_3 is small. Since it is relatively easy to know the values of M and l in industrial applications, we assume in this case that b_4 is known to be $b_4 = \dfrac{-1}{Ml}$, but $d_2(\mathbf{z})$ and $d_4(\mathbf{z})$ are still unknown. So, the controller (4-46) is implemented without u_r. With these modifications, the simulation results are shown in Figs. 6.8 to 11. It is seen in Fig. 6.8(a) that the tracking performance of the trolley is similar to the previous case. Fig. 6.8(b) shows that the rope angle after 20 seconds is improved which presents no oscillations. On the other hand, without the robust term, the control effort in Fig. 6.9 becomes smooth. Fig. 6.10 shows the trajectories of the error surfaces and Fig. 6.11 is the function approximation performance.

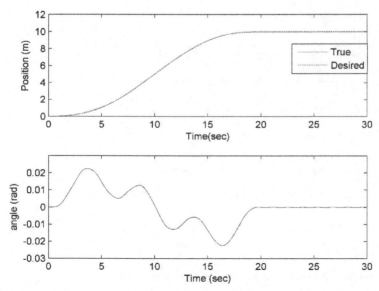

Figure 6.8.   System output with modified controller. (a) Trolley position (b) Rope angle

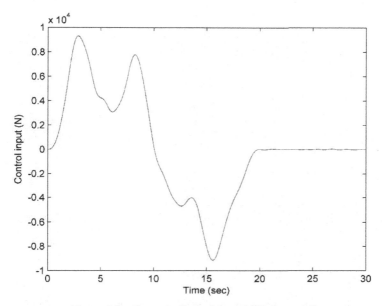

Figure 6.9. Control effort of the modified controller

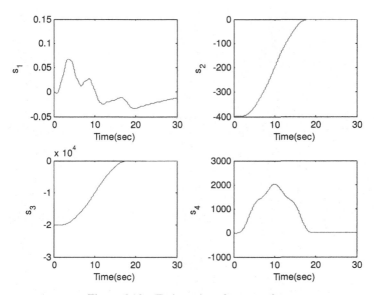

Figure 6.10. Trajectories of error surfaces

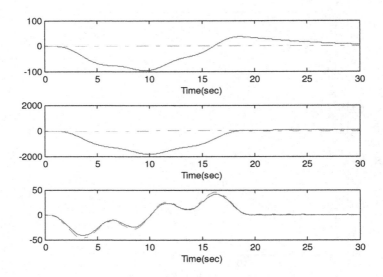

Figure 6.11. Function approximation performance (a) $\bar{d}_2$ (b) $\bar{d}_3$ (c) $\bar{d}_4$

Chapter 7

TORA System

The TORA (Translational Oscillator with Rotational Actuator) system is an undamped mechanical oscillator M controlled by the rotation of a pendulum represented by the mass m as shown in Fig. 7.1. It was firstly studied by Bernstein and co-workers (Wan et al., 1994; Bupp et al., 1994) and various control strategies were reported thereafter in the literature. Since the oscillator is undamped, the open-loop trajectory of M in response to a non-zero initial displacement will exhibit a sustained oscillation. Proper control of the pendulum motion may provide effective intervention to the oscillation. Most of the strategies are developed to regulate both the displacement x of the oscillator and the angle θ of the pendulum to the origin; therefore, it is an underactuated system.

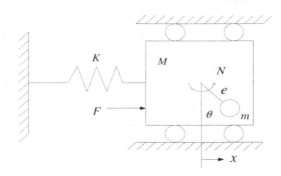

Figure 7.1. The TORA system

147

Jankovic et al. (1996) proposed cascade controllers and feedback passivating controllers to globally stabilize TORA systems asymptotically. Burg and Dawson (1997) suggested a pseudo-velocity filter to avoid the need for velocity feedback in the TORA controller developed in Jankovic et al. (1996). Tanaka et al. (1998) designed an optimal fuzzy controller based on LMI strategy for TORA control. Alleyne (1998) gave some physical insights on the passivity-based design for TORA systems suggested in Jankovic et al. (1996). Tadmor (2001) investigated the internal energy flow of the TORA system with respect to the performance of some control strategies. Petres et al. (2005) used tensor-product model transformation for affine decomposition of the TORA system. Lee et al. (2006) constructed a decoupled adaptive type-2 fuzzy control scheme to stabilize a TORA system. Hung et al. (2007) implemented a self-tuning fuzzy sliding controller for TORA systems. Nazrulla and Khalil (2008) designed a nonlinear output feedback controller for TORA system. Li et al. (2009) proposed a SIRMs-based type-2 fuzzy controller for regulating TORA dynamics. Chen and Huang (2010) decoupled the underactuated dynamics in TORA followed by the design of an adaptive controller to tolerate various internal uncertainties and external disturbances. Lee and Chang (2012) evaluated the adaptive backstepping control on TORA system experimentally.

In this chapter, the controller developed in chapter 4 is applied to the TORA system based on Chen and Huang (2010) so that the underactuated dynamics is transformed into a special cascade form followed by the design of an adaptive strategy. In section 7.1 the system dynamics of TORA is introduced. The transformation for decoupling the underactuated dynamics is derived in section 7.2. Simulation cases to verify efficacy of the controller are presented in section 7.3.

7.1 System Dynamics

The kinetic energy of the oscillator and the pendulum can respectively be represented to be

$$K_{oscillator} = \frac{1}{2}(M+m)\dot{x}^2$$

$$K_{pendulum} = \frac{1}{2}(I+me^2)\dot{\theta}^2 + me\dot{x}\dot{\theta}\cos\theta$$

where M and m are the masses of the oscillator and pendulum respectively, e is eccentricity of the pendulum, and I is the moment of inertial of the pendulum. The total kinetic energy is the combination of $K_{oscillator}$ and $K_{pendulum}$ as

$$K = K_{oscillator} + K_{pendulum}$$
$$= \frac{1}{2}(M+m)\dot{x}^2 + \frac{1}{2}(I+me^2)\dot{\theta}^2 + me\dot{x}\dot{\theta}\cos\theta \qquad (1)$$

With the definition of the potential energy of the system

$$V = -mg\cos\theta + \frac{1}{2}kx^2 \qquad (2)$$

the Lagrangian can then be found as

$$L = K - V$$
$$= \frac{1}{2}(M+m)\dot{x}^2 + \frac{1}{2}(I+me^2)\dot{\theta}^2 + me\dot{x}\dot{\theta}\cos\theta + mg\cos\theta - \frac{1}{2}kx^2 \qquad (3)$$

The generalized coordinate vector and the generalized force vector can respectively be assigned as

$$\mathbf{q} = \begin{bmatrix} q_1 \\ q_2 \end{bmatrix} = \begin{bmatrix} x \\ \theta \end{bmatrix} \in \Re^2$$

$$\mathbf{\tau} = \begin{bmatrix} F \\ N \end{bmatrix} \in \Re^2$$

where F is the disturbance acting on the oscillator and N is the control torque provided by a motor to drive the pendulum. Let us compute the quantities below to facilitate the derivation of the equation of motion via the Lagrange equation

$$\frac{\partial L}{\partial \dot{x}} = (M + m)\dot{x} + ml\dot{\theta}\cos\theta$$

$$\frac{\partial L}{\partial x} = -kx$$

$$\frac{\partial L}{\partial \dot{\theta}} = me\dot{x}\cos\theta + (I + me^2)\dot{\theta}$$

$$\frac{\partial L}{\partial \theta} = -me\dot{x}\dot{\theta}\sin\theta$$

then we may have the dynamic equation of the TORA system in the X-space using the Lagrange equation

$$\frac{d}{dt}\frac{\partial L}{\partial \dot{q}} - \frac{\partial L}{\partial q} = \tau$$

to have

$$(M + m)\ddot{x} + me\cos\theta\ddot{\theta} - me\dot{\theta}^2\sin\theta + kx = F$$
$$(me^2 + I)\ddot{\theta} + me\cos\theta\ddot{x} = N$$
(4)

Define the normalized state $p = \sqrt{\dfrac{M+m}{I+me^2}}\,x$, normalized time $\tau = \sqrt{\dfrac{k}{M+m}}\,t$, dimensionless control $u = \dfrac{M+m}{k(I+me^2)}N$ and dimensionless disturbance $w = \dfrac{1}{k}\sqrt{\dfrac{M+m}{I+me^2}}\,F$ (Petres et al., 2005), then (4) becomes

$$\ddot{p} + p = \varepsilon(\dot{\theta}^2\sin\theta - \ddot{\theta}\cos\theta) + w$$
$$\ddot{\theta} = -\varepsilon\ddot{p}\cos\theta + u$$
(5)

where the differentiations are with respect to the normalized time and ε represents the coupling between the translational and rotational motions and can be defined as

$$\varepsilon = \frac{me}{\sqrt{(I+me^2)(M+m)}} \tag{6}$$

Define the state vector

$$\mathbf{x} = [p \quad \dot{p} \quad \theta \quad \dot{\theta}]^T = [x_1 \quad x_2 \quad x_3 \quad x_4]^T \in \Re^4$$

and the state space representation for (5) becomes

$$\begin{aligned}
\dot{x}_1 &= x_2 \\
\dot{x}_2 &= f_2(\mathbf{x}) + b_2(\mathbf{x})u \\
\dot{x}_3 &= x_4 \\
\dot{x}_4 &= f_4(\mathbf{x}) + b_4(\mathbf{x})u
\end{aligned} \tag{7}$$

where

$$f_2(\mathbf{x}) = \frac{-x_1 + \varepsilon x_4^2 \sin x_3 + w}{1 - \varepsilon^2 \cos^2 x_3} \tag{8a}$$

$$b_2(x_3) = \frac{-\varepsilon \cos x_3}{1 - \varepsilon^2 \cos^2 x_3} \tag{8b}$$

$$f_4(\mathbf{x}) = \frac{\varepsilon x_1 \cos x_3 - \varepsilon^2 x_4^2 \sin x_3 - \varepsilon \cos x_3 w}{1 - \varepsilon^2 \cos^2 x_3} \tag{8c}$$

$$b_4(x_3) = \frac{1}{1 - \varepsilon^2 \cos^2 x_3} \tag{8d}$$

There is only one control u appears in both the dynamic equation of the oscillator and the pendulum in (7) and hence the whole system is underactuated.

7.2 Coordinate Transformation

It is apparently that (7) is in the standard form (4.41) with $n = 4$. We may then transform the system from the X-space into Z-space by using the mapping (4.42) as

$$
z_1 = x_1 - \int_0^{x_3} \frac{b_2(s)}{b_4(s)} ds = x_1 - \int_0^{x_3} \frac{\dfrac{-\varepsilon \cos x_3}{1-\varepsilon^2 \cos^2 x_3}}{\dfrac{1}{1-\varepsilon^2 \cos^2 x_3}} ds = x_1 + \varepsilon \sin x_3
$$

$$
z_2 = x_2 - \frac{b_2(\mathbf{x})}{b_4(\mathbf{x})} x_4 = x_2 - \frac{\dfrac{-\varepsilon \cos x_3}{1-\varepsilon^2 \cos^2 x_3}}{\dfrac{1}{1-\varepsilon^2 \cos^2 x_3}} = x_2 + \varepsilon x_4 \cos x_3
$$

$$
z_3 = x_3
$$

$$
z_4 = x_4
$$

(9)

With the transformation (9), the system dynamics in the Z-space becomes

$$
\begin{aligned}
\dot{z}_1 &= z_2 + d_1(\mathbf{z}) \\
\dot{z}_2 &= z_3 + d_2(\mathbf{z}) \\
\dot{z}_3 &= z_4 + d_3(\mathbf{z}) \\
\dot{z}_4 &= d_4(\mathbf{z}) + b_4(\mathbf{z})u
\end{aligned}
$$

(10)

where $d_1(\mathbf{z}) = d_3(\mathbf{z}) = 0$, and

$$
d_2(\mathbf{z}) = -z_3 - z_1 + \varepsilon \sin z_3 + w \tag{11a}
$$

$$
d_4(\mathbf{z}) = f_4(\mathbf{z}) = \frac{\varepsilon \cos z_3 (x_1 - \varepsilon z_4^2 \sin z_3) - \varepsilon \cos x_3 w}{1-\varepsilon^2 \cos^2 z_3} \tag{11b}
$$

$$
b_4(x_3) = \frac{1}{1-\varepsilon^2 \cos^2 x_3} \tag{11c}
$$

7.3 Simulation Cases

The system starts from $\mathbf{x}(0) = [1 \ \ 0 \ \ 1 \ \ 0]^T$, and the desired state is $\mathbf{x}_d = [0 \ \ 0 \ \ 0 \ \ 0]^T$, i.e., we would like to send the oscillator from 1 (m) back to the origin while the pendulum goes from 1 radian to the stable equilibrium at $\theta = 0$. The actual values of the system parameters are $m = 0.5$ (kg), $M = 2$ (kg), $I = 0.1$ (kgm^2) and $e = 0.5$ (m). By using (6) we may have $\varepsilon = 0.33$. The controller in (4-46) together with the desired trajectories (4-45) and the robust term (4-48) are used with the parameters selected as $c_1 = c_2 = 2$, $c_3 = 3$, and $c_4 = 0.3$. The first 21 terms of the Fourier series are used as the basis functions. The weighting matrices in (4-50) are picked as $\Gamma_i = 10\mathbf{I}_{21}$, $i = 1,...,4$. The constant σ in (4-50) is set to zero to turn off the σ-modification in the update law. The initial weighting in (4-50) are set to be $\hat{\mathbf{w}}_2(0) = [0.5 \quad 0 \quad \cdots \quad 0]^T \in \mathfrak{R}^{21}$ and $\hat{\mathbf{w}}_4(0) = [0.2 \quad 0 \quad \cdots \quad 0]^T \in \mathfrak{R}^{21}$. Two simulation cases are presented here. The first case is the control of an uncertain TORA. The second case presents the control result of an uncertain TORA with the presence of some sinusoidal disturbances on the oscillator and significant variation in system parameters.

CASE 1: Uncertain TORA

In this case, we regard d_2 and d_4 as time-varying uncertainties. In addition, b_4 is a bounded uncertainty whose bounds can be computed as $b_{\min} = 1 \leq b_4 \leq 1.122 = b_{\max}$ and further we may calculate $b_m = \sqrt{b_{\min}b_{\max}} = 1.059$, $\beta_{\min} = 0.944$ and $\beta_{\max} = 1.059$. The simulation results are shown in Figs. 7.2 to 7.5. Fig. 7.2 verifies that both the oscillator displacement and pendulum angle converge nicely regardless of the uncertainties in the system dynamics. Fig. 7.3 is the control effort required. Fig. 7.4 presents the convergence of each error surface. Fig. 7.5 shows the function approximation performance. It is seen that all parameters are bounded as desired.

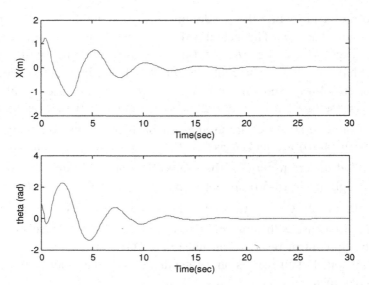

Figure 7.2.   Trajectories of (a) oscillator position (b) pendulum angle

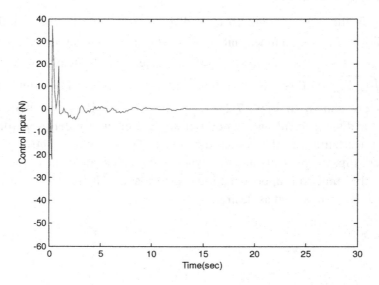

Figure 7.3.   Control effort

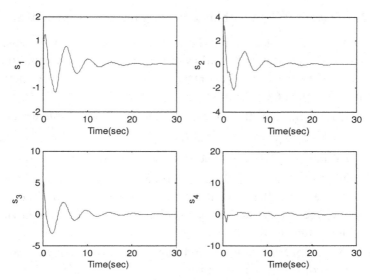

Figure 7.4. Trajectories of error surfaces

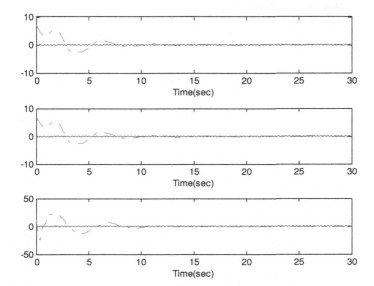

Figure 7.5. Function approximation performance (a) $\overline{d}_2$ (b) $\overline{d}_3$ (c) $\overline{d}_4$

CASE 2: Uncertain TORA with time-varying masses and disturbances

In this case, in addition to the system uncertainties, the masses of the oscillator and pendulum are allowed to vary with significant amount so that ε defined in (6) becomes $e = 0.33 + 0.2\sin t$. A periodic disturbance $F = 0.3\sin t$ is applied to the oscillator during $t \in [30,40]$ (sec) to test robustness of the strategy. The controller parameters can be computed to be $b_{\min} = 1 \le b_4 \le 1.391 = b_{\max}$, $b_m = \sqrt{b_{\min} b_{\max}} = 1.179$, $\beta_{\min} = 0.313$ and $\beta_{\max} = 0.848$. The simulation results are shown in Figs. 7.6 to 7.9. Fig. 7.6 shows that even with time-varying masses and various uncertainties the controller can still bring the system to the desired states. Application of the disturbance to the oscillator gives significant deviation of the system trajectories during $t \in [30,40]$. However, they are still bounded and go back to the desired states quickly when the external force is removed. Fig. 7.7 is the history of the control effort. Fig. 7.8 is the convergence of each error surface. Fig. 7.9 shows the function approximation performance.

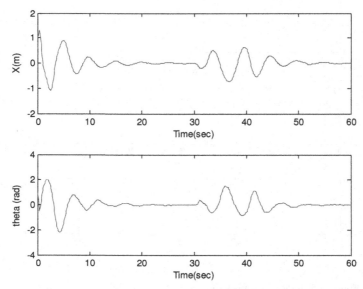

Figure 7.6.   Trajectories of (a) oscillator position (b) pendulum angle

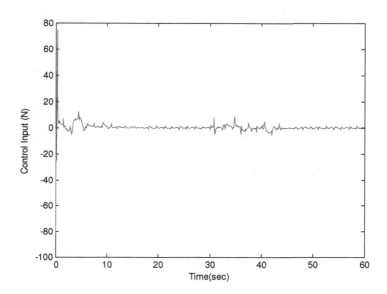

Figure 7.7.   Control effort

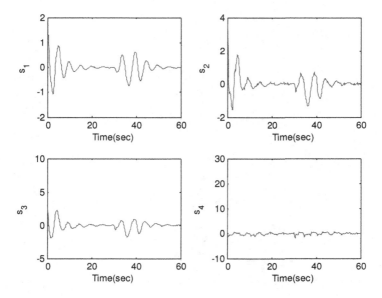

Figure 7.8.   Trajectories of error surfaces

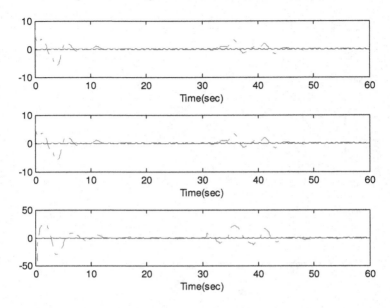

Figure 7.9.   Function approximation performance (a) $\bar{d}_2$ (b) $\bar{d}_3$ (c) $\bar{d}_4$

Chapter 8

Rotary Inverted Pendulum

The rotary inverted pendulum as shown in Fig. 8.1 was firstly introduced by Furuta (Furuta et al., 1991). The arm with length L is driven by the motor torque u. The rotation of the arm can then provide power to swing the pendulum which is similar to the traditional inverted pendulum. Since there is only one input to control both the arm and the pendulum, the rotary inverted pendulum is an underactuated system. Most of the controls for the rotary inverted pendulum fall into one of the several categories (Carlos et al., 2010). For example, some considered the problem of stabilizing the pendulum around the unstable vertical position (Carlos et al., 2010; Nair, 2002; Acosta, 2010; Olfati-Saber, 1999; Turker et al., 2012). Some swung the pendulum from its hanging position to its upright vertical position (Gordillo et al., 2003; Hera et al., 2009; Fantoni et al., 2002; Astrom et al., 2008; Sukontanakarn et al., 2009). Some others tried to create oscillations around its unstable vertical position (Shiriaev et al., 2007; Aguilar et al., 2009; Freidovich et al., 2009).

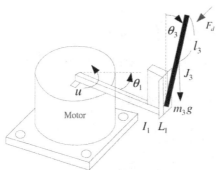

Figure 8.1. The rotary inverted pendulum

In this chapter, we would like to consider the control problem of stabilizing the pendulum around the unstable vertical position when subject to time-varying uncertainties. In section 8.1, the dynamic equation of the rotary inverted pendulum system is derived by using the Lagrange equation. The coordinate transformation introduced in chapter 3 is then performed in section 8.2 to represent the system in the Z-space. The last section gives the simulation cases to verify the effectiveness of the design.

8.1 System Dynamics

The kinetic energy of the arm and the pendulum can respectively be represented to be

$$K_{arm} = \frac{1}{2} I \dot{\theta}^2$$

$$K_{pendulum} = \frac{1}{2} J \dot{\phi}^2 + \frac{1}{2} m L^2 \dot{\theta}^2 + \frac{1}{2} m l^2 \dot{\phi}^2 + \frac{1}{2} m l^2 \sin^2 \phi \dot{\theta}^2 + m L l \dot{\theta} \dot{\phi} \cos \phi$$

where m and J are respectively the mass and the moment of inertial of the pendulum, l is the distance of the center of mass of the pendulum to the tip point, I is the moment of inertial of the arm, θ is the displacement of the arm and ϕ is the displacement of the pendulum. The total kinetic energy is the combination of K_{arm} and $K_{pendulum}$ as

$$\begin{aligned} K = K_{arm} &+ K_{pendulum} \\ &= \frac{1}{2} I \dot{\theta}^2 + \frac{1}{2} J \dot{\phi}^2 + \frac{1}{2} m L^2 \dot{\theta}^2 + \frac{1}{2} m l^2 \dot{\phi}^2 \\ &+ \frac{1}{2} m l^2 \sin^2 \phi \dot{\theta}^2 + m L l \dot{\theta} \dot{\phi} \cos \phi \end{aligned} \tag{1}$$

With the definition of the potential energy of the pendulum

$$V = mgl(\cos \theta - 1) \tag{2}$$

the Lagrangian can then be found as

$$L = K - V$$

$$= \frac{1}{2}I\dot{\theta}^2 + \frac{1}{2}J\dot{\phi}^2 + \frac{1}{2}mL^2\dot{\theta}^2 + \frac{1}{2}ml^2\dot{\phi}^2 \tag{3}$$

$$+ \frac{1}{2}ml^2\sin^2\phi\dot{\theta}^2 + mLl\dot{\theta}\dot{\phi}\cos\phi - mgl(\cos\theta - 1)$$

The generalized coordinate vector and the generalized force vector can respectively be assigned as

$$\mathbf{q} = \begin{bmatrix} q_1 \\ q_2 \end{bmatrix} = \begin{bmatrix} \theta \\ \phi \end{bmatrix} \in \Re^2$$

$$\boldsymbol{\tau} = \begin{bmatrix} u \\ F_d \end{bmatrix} \in \Re^2$$

where F_d is the external disturbance acting on the pendulum. Let us compute the quantities below to facilitate the derivation of the equation of motion via the Lagrange equation.

$$\frac{\partial L}{\partial \dot{\theta}} = [I + m(L^2 + l^2\sin^2\phi)]\dot{\theta} + mlL\cos\phi\dot{\phi}$$

$$\frac{\partial L}{\partial \theta} = 0$$

$$\frac{\partial L}{\partial \dot{\phi}} = mlL\dot{\theta}\cos\phi + (J + ml^2)\dot{\phi} \tag{4}$$

$$\frac{\partial L}{\partial \phi} = ml^2\sin\phi\cos\phi\dot{\theta}^2 - mlL\dot{\theta}\dot{\phi}\sin\phi + mgl\sin\phi$$

then we may have the dynamic equation of the rotary inverted pendulum system in the X-space using the Lagrange equation

$$\frac{d}{dt}\frac{\partial L}{\partial \dot{\mathbf{q}}} - \frac{\partial L}{\partial \mathbf{q}} = \boldsymbol{\tau} \tag{5}$$

to have

$$\ddot{\theta} = \frac{1}{\Delta}[(ml^2 + J)u - (ml^2 + J)ml^2 \sin(2\phi)\dot{\theta}\dot{\phi}$$

$$-\tfrac{1}{2}m^2l^2L\cos\phi\sin(2\phi)\dot{\theta}^2 + (ml^2 + J)mlL\sin\phi\dot{\phi}^2 \qquad (6a)$$

$$-m^2l^2Lg\sin\phi\cos\phi - mlL\cos x_3 F_d]$$

$$\ddot{\phi} = \frac{1}{\Delta}\{-(mlL\cos\phi)u - m^2l^2L^2\sin\phi\dot{\phi}$$

$$+ml^2\sin(2\phi)\dot{\theta}[mlL\cos\phi\dot{\phi} + \tfrac{1}{2}(I + mL^2 + ml^2\sin^2\phi)\dot{\theta}] \qquad (6b)$$

$$+(I + mL^2 + ml^2\sin^2\phi)mgl\sin\phi + (I + mL^2 + ml^2\sin^2 x_3)F_d\}$$

where $\Delta = (J + ml^2)(I + ml^2\sin^2\phi) + JmL^2 + m^2l^2L^2\sin^2\phi$. Define the state vector

$$\mathbf{x} = [\theta \quad \dot{\theta} \quad \phi \quad \dot{\phi}]^T = [x_1 \quad x_2 \quad x_3 \quad x_4]^T \in \mathfrak{R}^4$$

and the state space representation becomes

$$\begin{aligned}
\dot{x}_1 &= x_2 \\
\dot{x}_2 &= f_2(\mathbf{x}) + b_2(\mathbf{x})u \\
\dot{x}_3 &= x_4 \\
\dot{x}_4 &= f_4(\mathbf{x}) + b_4(\mathbf{x})u
\end{aligned} \qquad (7)$$

where

$$f_2(\mathbf{x}) = \frac{(J + ml^2)mlx_4[Lx_4\sin x_3 - lx_2\sin(2x_3)]}{\Delta}$$

$$-\frac{m^2l^2L\cos x_3[g\sin x_3 + 0.5lx_2^2\sin(2x_3)]}{\Delta} - \frac{mlL\cos x_3}{\Delta}F_d \qquad (8a)$$

$$b_2(x_3) = \frac{J + ml^2}{\Delta} \qquad (8b)$$

$$f_4(\mathbf{x}) = \frac{ml \sin x_3 [(I + mL^2 l \sin^2 x_3) g - mlL^2 x_3^2 \cos x_3]}{\Delta}$$

$$- \frac{ml^2 x_2 \sin(2x_3) x_2 [mlL x_4 \cos x_3 + 0.5 x_2 (I + mL^2 + l^2 \sin^2 x_3)]}{\Delta}$$

$$+ \frac{I + mL^2 + ml^2 \sin^2 x_3}{\Delta} F_d \tag{8c}$$

$$b_2(x_3) = \frac{-mlL \cos x_3}{\Delta} \tag{8d}$$

There is only one control u appears in both the dynamic equations of the arm and the pendulum in (7) and hence the whole system is underactuated.

8.2 Coordinate Transformation

It is apparently that (7) is in the standard form (4.41) with $n = 4$. We may then transform the system from the X-space into Z-space by using the mapping (4.42) as

$$z_1 = x_1 - \int_0^{x_3} \frac{b_2(s)}{b_4(s)} ds$$

$$= x_1 - \int_0^{x_3} \frac{\dfrac{J + ml^2}{\Delta}}{\dfrac{-mlL \cos s}{\Delta}} ds = x_1 + \frac{J + ml^2}{mlL} \ln|\sec x_3 + \tan x_3| \tag{9a}$$

$$z_2 = x_2 - \frac{b_2(\mathbf{x})}{b_4(\mathbf{x})} x_4 = x_2 - \frac{\dfrac{J + ml^2}{\Delta}}{\dfrac{-mlL \cos x_3}{\Delta}} = x_2 + \frac{J + ml^2}{mlL} x_4 \sec x_3 \tag{9b}$$

$$z_3 = x_3 \tag{9c}$$

$$z_4 = x_4 \tag{9d}$$

With the transformation (9), the system dynamics in the Z-space becomes

$$\dot{z}_1 = z_2 + d_1(\mathbf{z})$$
$$\dot{z}_2 = z_3 + d_2(\mathbf{z})$$
$$\dot{z}_3 = z_4 + d_3(\mathbf{z})$$
$$\dot{z}_4 = d_4(\mathbf{z}) + b_4(\mathbf{z})u$$

(10)

where $d_1(\mathbf{z}) = d_3(\mathbf{z}) = 0$, and

$$d_2(\mathbf{z}) = -z_3 - f_2(\mathbf{z}) + \frac{J + ml^2}{mlL} f_4(\mathbf{z}) \sec x_3$$

(11a)

$$d_4(\mathbf{z}) = f_4(\mathbf{z})$$

(11b)

$$b_4(z_3) = \frac{-mlL \cos z_3}{\Delta}$$

(11c)

It is noted that both $d_2(\mathbf{z})$ and $d_4(\mathbf{z})$ in (11) are extremely complex due to the complex forms of $f_2(\mathbf{z})$ and $f_4(\mathbf{z})$ in (8) under the transformation in (9). Here, we would like to assume that both $d_2(\mathbf{z})$ and $d_4(\mathbf{z})$ are time-varying uncertainties and $b_4(z_3)$ is an uncertainty with known bound.

8.3 Simulation Cases

To verify the effectiveness of the controller designed in chapter 4, let us consider the rotary inverted pendulum system in Fig. 8.1. The initial condition is assumed to be $\mathbf{x}(0) = [1 \quad 0 \quad -1 \quad 0]^T$ and we would like to bring the system state to the origin. The actual values of the system parameters (Fantoni et al., 2001) are $m = 5.38 \times 10^{-2}$ (kg), $J = 1.98 \times 10^{-4}$ (kgm^2), $l = 0.113$ (m), $L = 0.215$ (m) and $I = 1.75 \times 10^{-2}$ (kgm^2). The parameters for modeling b_4 can be computed to be $b_{max} = 81.79$, $b_{min} = 40.08$, $b_m = 57.26$, $\beta_{max} = 1.4284$ and $\beta_{min} = 0.700$. The controller in (4-46) together with the desired trajectories (4-45) and the robust term (4-48) are used. The weighting matrices in (4-50) are picked as $\Gamma_1 = \Gamma_2 = 32\mathbf{I}_{21}$, $\Gamma_3 = 35\mathbf{I}_{21}$, and $\Gamma_4 = 40\mathbf{I}_{21}$. The constant σ in (4-50) is set to zero to turn off the σ-modification in the update law. The first 21 terms of the Fourier series are selected as the basis functions. The initial weighting in (4-50) are set to be $\hat{\mathbf{w}}_i(0) = [0 \quad 0 \quad \cdots \quad 0]^T \in \mathfrak{R}^{21}$ for all $i = 1,\dots,4$. A periodic disturbance $F_d = \cos 2t$ is applied to the cart during $t \in [4,6]$ (sec) to test robustness of the strategy. Two simulation cases are presented here. The first case is the control of an uncertain rotary inverted pendulum. The second case presents the control result of an uncertain rotary inverted pendulum with the presence of significant variation in system parameter.

CASE 1: Uncertain inverted pendulum

In this case the controller parameters are selected as $c_1 = 50$, $c_2 = 20$, $c_3 = 0.06$, and $c_4 = 1 \times 10^{-9}$. The simulation results are shown in Figs. 8.2 to 8.5. Fig. 8.2 verifies that both the arm displacement and pendulum angle converge nicely regardless of the uncertainties in the system dynamics. Application of the disturbance to the pendulum gives significant deviation of the system trajectories during $t \in [4,6]$. However, they are still bounded and go back to the desired states quickly when the external force is removed. Fig. 8.3 is the control effort required. Fig. 8.4 presents the convergence of each error surface. Fig. 8.5 hows the function approximation performance. It is seen that all parameters are bounded as desired.

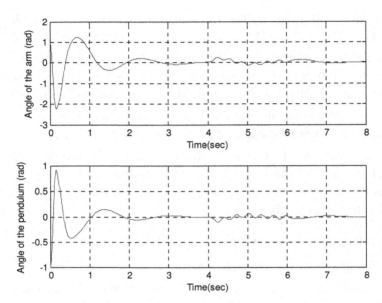

Figure 8.2.   Trajectories of (a) arm position (b) pendulum angle

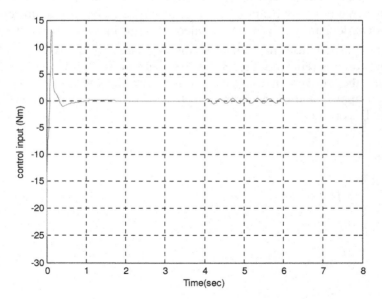

Figure 8.3.   Control effort

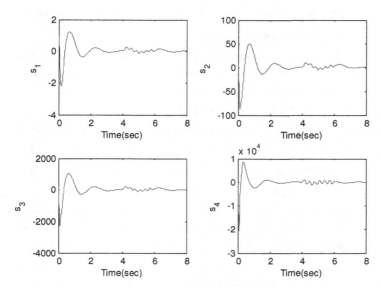

Figure 8.4. Trajectories of error surfaces

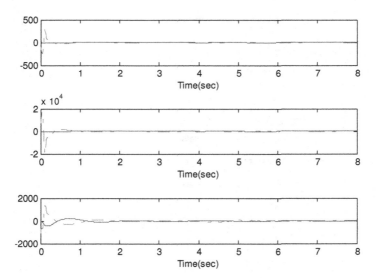

Figure 8.5. Function approximation performance (a) $\bar{d}_2$ (b) $\bar{d}_3$ (c) $\bar{d}_4$

CASE 2: Uncertain inverted pendulum with time-varying parameters

In this case, in addition to the system uncertainties and external disturbances, the mass and the length of the pendulum are allowed to have rapid variation with significant amount as $m = 0.0538 + 0.04\cos 50t$ (kg) and $l = 0.113 + 0.1\sin 4t$ (m). The controller parameters can be computed to be $b_{min} = 31.2$, $b_{max} = 534.95$, $b_m = 129.2$, $\beta_{max} = 4.14$ and $\beta_{min} = 0.2415$. Other controller parameters are selected as $c_1 = 60$, $c_2 = 30$, $c_3 = 0.06$, $c_4 = 1 \times 10^{-8}$, $\Gamma_1 = \Gamma_2 = 32\mathbf{I}_{21}$, $\Gamma_3 = 35\mathbf{I}_{21}$ and $\Gamma_4 = 40\mathbf{I}_{21}$. The simulation results are shown in Figs. 8.6 to 8.9. Fig. 8.6 shows that even with time-varying parameters and various uncertainties the controller can still bring the system to the desired states. Application of the disturbance to the cart gives significant deviation of the system trajectories during $t \in [4,6]$. However, they are still bounded and go back to the desired states quickly when the external force is removed. Fig. 8.7 is the history of the control effort. Fig. 8.8 is the convergence of each error surface. Fig. 8.9 shows the function approximation performance.

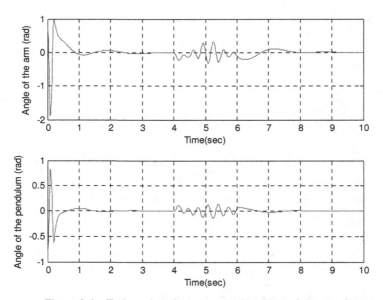

Figure 8.6.   Trajectories of (a) arm position (b) pendulum angle

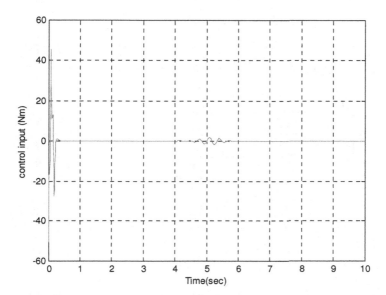

Figure 8.7. Control effort

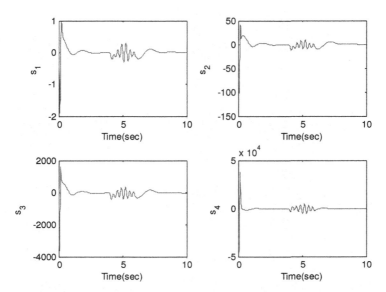

Figure 8.8. Trajectories of error surfaces

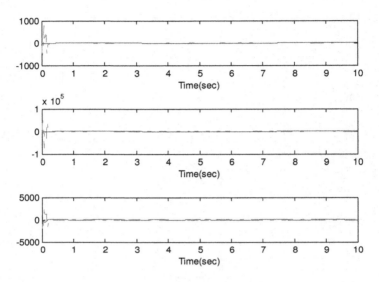

Figure 8.9. Function approximation performance (a) $\overline{d}_2$ (b) $\overline{d}_3$ (c) $\overline{d}_4$

Chapter 9

Vibration Absorber

In this chapter, we would like to consider the vibration suppression of a primary system which is modeled as a one-dimensional mechanical system composed of impedance elements as shown in Fig. 9.1. The primary system exhibits a sustained oscillation due to the external excitation f.

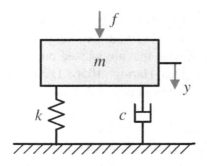

Figure 9.1. A primary system

    To reduce the vibration in the primary system, several approaches have been proposed. The simplest one is the passive vibration absorber (Ormondroyd and Den Hartog, 1928; Semercigil et al., 1992; Sun et al., 1995; Rana and Soong, 1998; Filipović and Schröder, 1998; Wright et al., 2004) which is also constructed with mechanical impedance elements (Fig. 9.2).

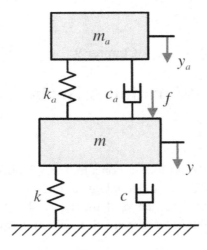

Figure 9.2.   A primary system with a passive vibration absorber

The vibration absorber in Fig. 9.2 changes the dynamics of the primary system in such a way that the original resonance frequency is replaced with two new ones that are located on both sides of the original resonance frequency (Den Hartog, 1956; Davis and Lesieutre, 2000). This can be seen in the frequency responses in Fig. 9.3 where there is only one resonance frequency ω_0 when the absorber is absent (solid line). The frequency ω_0 becomes a valley in the new spectrum (dotted line) due to the presence of the passive absorber; therefore effective vibration suppression can be achieved for disturbances near ω_0. However, if the disturbance frequency variation is large enough, either one of the two new resonance frequencies ω_{n1} and ω_{n2} might be excited, and the passive design fails. In other words, the passive vibration absorber is only effective in a neighborhood of a single frequency.

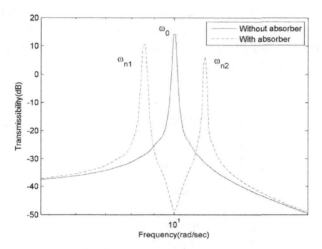

Figure 9.3. Frequency response with and without passive absorber. When without the absorber, the frequency response curve contains one peak. It is attenuated significantly when the absorber is installed, while two additional peaks are created. If the external excitation frequency is fixed, then the absorber is feasible to give satisfactory performance. However, if the excitation frequency deviates from the nominal value, the two additional peaks may be excited and the absorber fails.

To have an effective vibration absorber for primary systems with excitation frequencies varying in a significant amount, some tunable mechanical impedance is needed for changing the resonance frequency of the system. Several semi-active (Walsh and Lamancusa, 1992; Hollkamp and Starchville, 1994; Abé and Igusa, 1996; Nagaya et al., 1999; Brennan and Dayou, 2000; Williams et al., 2002; Koo et al., 2004) and active (Olgac and Holm-Hansen, 1994; Huang and Lian, 1994; Olgac and Jalili, 1998; Burdisso and Heilmann, 1998; Oueini et al., 1998; Filipovic and Schroder, 1999; Hosek et al., 1999; Olgac and Elmali, 2000; Bupp et al., 2000; Oueini and Nayfeh, 2000; Morgan and Wang, 2002; Cohen et al., 2002; Jalili1 and Knowles IV, 2004; Chen et al., 2005; Li et al., 2007; El-Badawy and El-Deen, 2007; Huyanan and Sims, 2007; Sun et al., 2007) designs were proposed to support automatic impedance adjustments. However, these adjustments depend on the knowledge of the excitation spectrums. If the excitation contains uncertainties, then these designs are not applicable. In addition to the single frequency disturbances, the primary system might subject to the broadband excitations, and the mechanical impedance in the vibration absorber tuned specifically to the tonal excitation can no longer give sufficient compensation which might largely deteriorate the system performance.

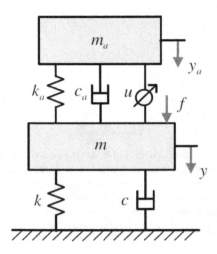

Figure 9.4. A primary system with an active vibration absorber

In this chapter, we would like to design a vibration absorber to a 1-DOF primary system whose external disturbance spectrum is not available. Since the excitation frequency may not be constant, the traditional passive design is not suggested. Because the disturbance spectrum is not given, most of the semi-active or active designs are not feasible. In addition, most of the actuators of active absorbers are installed between the primary system and the passive absorber (Fig. 9.4) resulting in an underactuated dynamics. Therefore, we are going to use the transformation in chapter 3 to decouple the system followed by applying the controller developed in chapter 4.

9.1 System Dynamics

Consider the active vibration absorber shown in Fig. 9.4. The primary system is modeled as a mass-spring-damper system with parameters m, k and c. The vibration absorber is constructed by attaching a reaction mass m_a to the primary system through spring k_a and damper c_a. The control force u is acting between the absorber and the primary system. The excitation force f is assumed to be unknown. Let y and y_a be the displacement of the primary system and absorber respectively, then the equations of motion of the combined system can be represented as

$$m\ddot{y} = -c\dot{y} - ky - c_a(\dot{y} - \dot{y}_a) - k_a(y - y_a) + f + u \tag{1a}$$

$$m_a\ddot{y}_a = c_a(\dot{y} - \dot{y}_a) + k_a(y - y_a) - u \tag{1b}$$

Define $\mathbf{x} = [x_1 \quad x_2 \quad x_3 \quad x_4]^T = [y \quad \dot{y} \quad y_a \quad \dot{y}_a]^T \in \Re^4$ to be the state vector in the X-space, and then we may rewrite equation (1) into the state space form

$$\begin{aligned}
\dot{x}_1 &= x_2 \\
\dot{x}_2 &= f_2(\mathbf{x}) + b_2(\mathbf{x})u \\
\dot{x}_3 &= x_4 \\
\dot{x}_4 &= f_4(\mathbf{x}) + b_4(\mathbf{x})u
\end{aligned} \tag{2}$$

where

$$f_2(\mathbf{x}) = \frac{1}{m}[-cx_2 - kx_1 - c_a(x_2 - x_4) - k_a(x_1 - x_3) + f]$$

$$b_2 = \frac{1}{m}$$

$$f_4(\mathbf{x}) = \frac{1}{m_a}[c_a(x_2 - x_4) + k_a(x_1 - x_3)]$$

$$b_4 = -\frac{1}{m_a}$$

It can be seen that the control force u is shared by the two subsystems. Hence, the whole system is underactuated. This implies that it is very difficult to design a controller to ensure that the vibration in the primary system is reduced and the amplitude of the absorber motion is limited in a specified range in the meantime.

9.2   Coordinate Transformation

It is apparently that (2) is in the standard form (4.41) with $n = 4$. We may then transform the system from the X-space into Z-space by using the mapping (4.42) as

$$z_1 = x_1 - \int_0^{x_3} \frac{b_2(s)}{b_4(s)} ds = x_1 - \int_0^{x_3} \frac{\dfrac{1}{m}}{-\dfrac{1}{m_a}} ds = x_1 + \frac{m_a}{m} x_3$$

$$z_2 = x_2 - \frac{b_2(\mathbf{x})}{b_4(\mathbf{x})} x_4 = x_2 - \frac{\dfrac{1}{m}}{-\dfrac{1}{m_a}} x_4 = x_2 + \frac{m_a}{m} x_4 \tag{3}$$

$$z_3 = x_3$$

$$z_4 = x_4$$

where $\mathbf{z} = [z_1 \quad z_2 \quad z_3 \quad z_4]^T \in \mathfrak{R}^4$ is the new state vector in the Z-space. The system can then be represented in the new space as

$$\dot{z}_1 = z_2 + d_1(\mathbf{z})$$
$$\dot{z}_2 = z_3 + d_2(\mathbf{z})$$
$$\dot{z}_3 = z_4 + d_3(\mathbf{z}) \tag{4}$$
$$\dot{z}_4 = d_4(\mathbf{z}) + b_4(\mathbf{z})u$$

where $d_1(\mathbf{z}) = d_3(\mathbf{z}) = 0$, and

$$d_2(\mathbf{z}) = \frac{1}{m}\left\{-c\left(z_2 - \frac{m_a}{m}z_4\right) - k\left(z_1 - \frac{m_a}{m}z_3\right) - c_a\left[\left(z_2 - \frac{m_a}{m}z_4\right) - z_4\right]\right.$$

$$\left. - k_a\left[\left(z_1 - \frac{m_a}{m}z_3\right) - z_3\right] + f\right\} - \frac{b_2}{b_4}\frac{1}{m_a}\left\{\left[\left(z_2 - \frac{m_a}{m}z_4\right) - z_4\right]\right.$$

$$\left. + k_a\left[\left(z_1 - \frac{m_a}{m}z_3\right) - z_3\right]\right\} - z_3$$

$$d_4(\mathbf{z}) = \frac{1}{m_a}\left\{c_a\left[\left(z_2 - \frac{m_a}{m}z_4\right) - z_4\right] + k_a\left[\left(z_1 - \frac{m_a}{m}z_3\right) - z_3\right]\right\}$$

9.3 Steady State Performance Analysis

In chapter 4, we have proved convergence of the states in the Z-space which, however, does not imply convergence in the X-space. This can be seen from the transformation (3a) as

$$z_1 = x_1 + \frac{m_1}{m}x_3 = y + \frac{m_a}{m}y_a \tag{5}$$

In the steady state, $z_1 \approx z_d = 0$ and we may write (5) as

$$y = -\frac{m_a}{m}y_a \tag{6}$$

Hence, the displacement of the primary system is proportional to that of the absorber, and we may not conclude convergence of y. To have further investigation, some straightforward derivation of (4b) under steady state assumption will give

$$y_a = -\frac{m}{km_a}f \tag{7}$$

Since both m and k cannot be altered, the only possible way to reduce y_a is via increasing of m_a. On the other hand, by substituting (7) into (6), we have

$$y = \frac{1}{k} f \tag{8}$$

Hence, the current design may not provide effective vibration suppression. However, let us look at (5) again. If the desired output is assigned as $z_{1d} = \beta y$ where $\beta \neq 1$ is to be selected. Then, under the steady state assumption, we may have

$$y = \frac{m_a}{m(\beta - 1)} y_a \tag{9}$$

Therefore, by proper adjustment of β, we may reduce the displacement of the primary system significantly.

9.4 Simulation Cases

In this section, four simulation cases are performed to justify the effectiveness of the strategy designed in chapter 4. The first case is the impact suppression that verifies the transient performance of the proposed controller. The second case is the single frequency excitation suppression. In the third case, the excitation force is considered as a harmonic signal with varying frequencies. In the last case, a broadband excitation is employed to verify the proposed controller. In all simulation cases, the excitation spectrums are assumed to be unavailable. The values of the primary system and vibration absorber parameters are: $m = 1$ (kg), $c = 2$ (Ns/m), $k = 100$ (N/m), $m_a = 0.1$ (kg), $c_a = 0.1$ (Ns/m), and $k_a = 10$ (N/m). The primary system has a resonant frequency at $\omega_0 = 10$ (rad/sec), and the vibration absorber eliminates this resonance but introduces two new ones at $\omega_{n1} = 8.54$ (rad/sec) and $\omega_{n2} = 11.7$ (rad/sec). The controller parameters are selected as $c_1 = 10$, $c_2 = 50$, $c_3 = 100$, and $c_4 = 200$. The first 21 terms of the Fourier series are employed as the basis functions. The adaptive gain matrices are selected as $\Gamma_i = 100\mathbf{I}$, $i = 1,...,4$.

CASE 1: Impact excitation

In this case, an impact force is applied to the primary system in the form:

$$f(t) = \begin{cases} 1 & \text{for } t \in [1, 1.1), \\ 0 & \text{else.} \end{cases} \tag{10}$$

The desired trajectory is $x_{1d} = 0$. The simulation results are shown in Figs. 9.5 to 9.8. Fig. 9.5 shows that it takes 6 seconds for the primary system to decay to close to zero when there is no absorber. Fig. 9.6 presents the performance of the proposed active absorber. It can be observed that the transient phase vanishes within 0.2 seconds and the steady state stays at zero without deviation. This result shows that the proposed strategy gives obvious improvement of vibration suppression when the system is subject to impact excitations. Fig. 9.7 is the control signal which shows that the algorithm is realizable. The time histories of the internal error signals of the adaptive controller are shown in Fig. 9.8, and it can be seen that they are all bounded as desired.

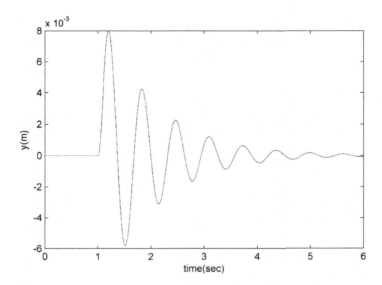

Figure 9.5. Impact response of the primary system without absorber in case 1

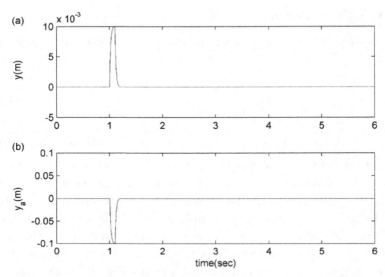

Figure 9.6.   Impact response of the primary system with active absorber in Case 1: (a) primary system and (b) active absorber

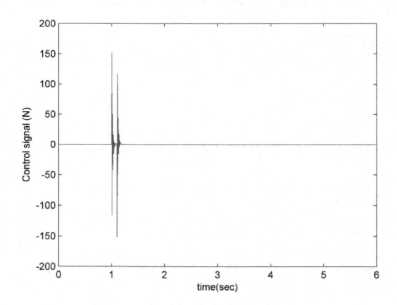

Figure 9.7.   Control signal in case 1

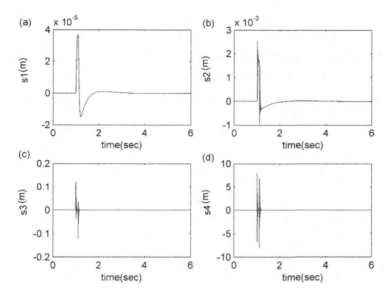

Figure 9.8. Convergence of error surfaces

CASE 2: Single-frequency excitation

The excitation in this case is a single-frequency signal $f = \sin 8.5t$ whose frequency is specifically selected to be exactly the resonance frequency of the primary system. The desired trajectory here is still picked as $z_{1d} = 0$. Fig. 9.9 presents the performance of 3 designs where Fig. 9.9(a) is the response of the primary system without any absorber, Fig. 9.9(b) is with the passive absorber, and Fig. 9.9(c) is with the proposed active absorber. When without any absorber, there will be a 3 (cm) displacement for the primary system, while for the passive design the displacement is increased to 4.5 (cm) even when the excitation frequency is so near its optimized frequency. The proposed active absorber can limit the displacement to be within 1 (cm). This result conforms to the steady-state performance (8) for the proposed method. Besides, equation (7) is able to be verified by using Fig. 9.10 where larger m_a implies smaller amplitude of y_a, but y is independent to the variation of m_a. To give more suppression of the primary system amplitude, we select $z_{1d} = \beta y$, and (9) is justified in Fig. 9.11 with 3 different values for β. Fig. 9.11(a) is the same as Fig. 9.9(c) where $\beta = 0$ and a 1 (cm) amplitude is obtained. When $\beta = 5$, the amplitude is reduced to 0.2 (cm); while when $\beta = 10$, it is further limited to be within 0.1 (cm). Therefore, by proper selection of β, we may ensure the amplitude of the primary system to be within some prescribed values.

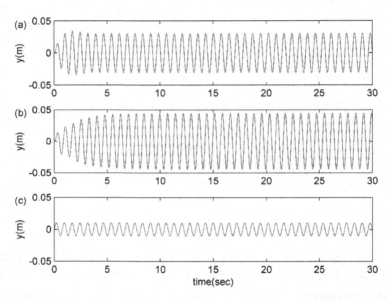

Figure 9.9. Response of the primary system with single-frequency excitation in Case 2: (a) no absorber, (b) passive absorber, (c) active absorber. It is seen that without the absorber the primary system exhibits an oscillation with 3 (cm) displacement. The second plot shows that the passive absorber may not give satisfactory performance when the excitation is near its resonance frequency. The proposed design limits the displacement to be within 1 (cm) which confirms the steady state performance in (8).

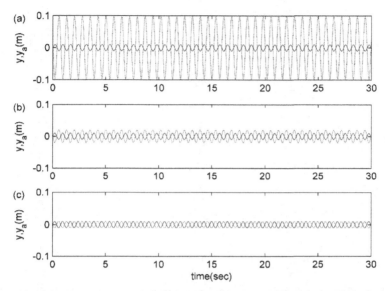

Figure 9.10. Vibration control performance of the active absorber under different m_a in Case 2: (a) $m_a = 0.1$ kg, (b) $m_a = 0.5$ kg, (c) $m_a = 1.0$ kg. The solid line is the primary system displacement y and dotted line is the vibration absorber displacement y_a. This result indicates that heavier m_a can effectively reduce y_a amplitude and hence equation (7) is explicitly verified.

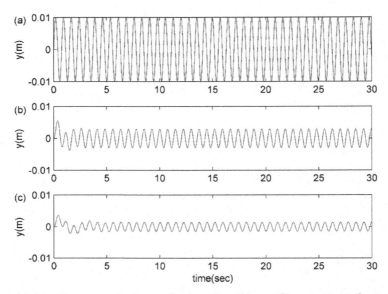

Figure 9.11.   Response of primary system under different β in case 2: (a) $\beta = 0$, (b) $\beta = 5$, (c) $\beta = 10$. This result shows that with larger β the primary system vibration can be significantly reduced.

CASE 3: Frequency-varying excitations

The excitation considered in this case is a frequency-varying sinusoidal signal which has the form:

$$f(t) = \begin{cases} \sin 8.54t & \text{for } t \in [0, 10), \\ \sin 10t & \text{for } t \in [10, 20), \\ \sin 11.7t & \text{for } t \in [20, 30) \end{cases} \quad (11)$$

The desired trajectory is $z_{1d} = 0$. In this case, the frequency variation covers the first resonance frequency $\omega_{n1} = 8.54$ (rad/sec), original resonance frequency $\omega_0 = 10$ (rad/sec), and second resonance frequency $\omega_{n2} = 11.7$ (rad/sec). The simulation results are shown in Figs. 9.12 to 9.15. Fig. 9.12 presents the performance of 3 designs where Fig. 9.12(a) is the response of the primary system without any absorber, 12(b) is with the passive absorber, and 12(c) is with the proposed active absorber.

When without any absorber, different amplitudes of the primary system give responses to different excitation frequencies. When the passive absorber is used, it is only effective for vibration suppression within $t \in [10, 20)$, because other frequencies are outside the valid range for its design. However, Fig. 9.12(c) shows that the proposed active absorber can deal with all excitation frequencies and results in a uniform small amplitude in the primary system. Fig. 9.13 is the control signal. Fig. 9.14 shows the amplitude of the primary system under different β. This result reconfirms the effectiveness of β adjustment in vibration amplitude suppression. The transmissibility curves of the system are shown in Fig. 9.15. It is seen that the proposed controller provides a uniform low value near the resonance frequency and hence can give almost identical output amplitude regardless of the variation of the excitation frequency.

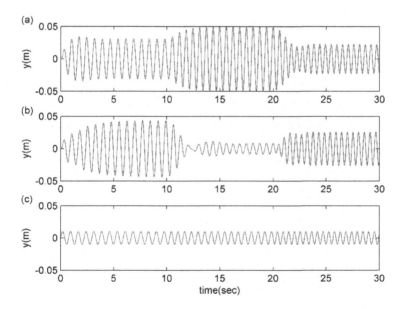

Figure 9.12. Response of the primary system in case 3: (a) no absorber (b) passive absorber (c) active absorber. It is seen that the passive absorber is only valid near 15 seconds, while the proposed design can reduce the vibration for all given excitation frequencies.

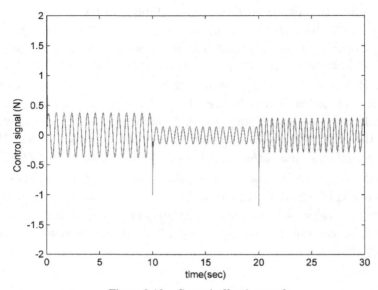

Figure 9.13.   Control effort in case 3

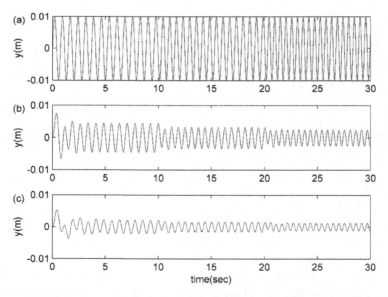

Figure 9.14.   Response of primary system under different β in case 3: (a) β=0 (b) β=5 (c) β=10. The larger the value of β, the smaller the output displacement.

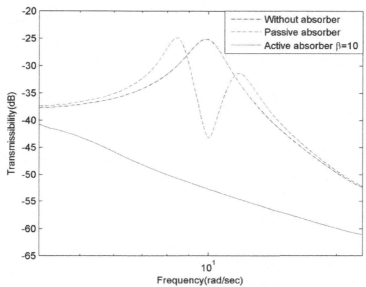

Figure 9.15. Frequency response comparison in case 3. The proposed design with the selection of $\beta = 10$ provides a uniform low value near the original resonance frequency which implies almost identical output amplitude regardless of the variation of the excitation frequency.

CASE 4: Broadband excitations

To test the performance of the design under broadband excitation, the signal $f(t) = \cos 4t + \sin 8.54t + \cos 10t + \sin 11.7t + \cos 16t + \sin t^2$ is used. The desired trajectory is designed as $z_{1d} = \beta y$ $z_{1d} = \beta y$ with $\beta = 100$. Fig. 9.16 presents the performance of 3 designs where Fig. 9.16(a) is the response of the primary system without any absorber, 16(b) is with the passive absorber, and 16(c) is with the proposed active absorber. It is seen that the passive absorber can no longer give satisfactory vibration suppression when the broadband excitation is applied. However, for the proposed controller with sufficient large value of β, the amplitude of the primary system is able to be largely reduced to 0.1 (cm). A further simulation result is shown in Fig. 9.17 in terms of the transmissibility curves. We see that a large β enables the proposed controller to push the curve down below -55 (dB) in a wide frequency range uniformly.

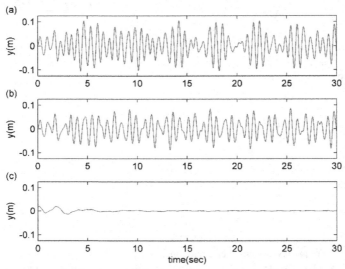

Figure 9.16.   Response of the primary system in case 4: (a) no absorber (b) passive absorber (c) active absorber with $\beta=100$.

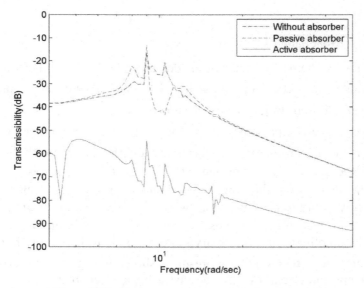

Figure 9.17.   Frequency response comparison in case 4

Chapter 10

Pendubot

A pendubot is a 2-link robot arm (Fig. 10.1) whose first link is driven by a motor while the second link is not actuated. Since there is only one control effort to regulate the motion of the two links, it is underactuated. Most of the control problems are to swing up the two links to the upright position. Spong and Block (1995) described the dynamics and control of a pendubot. Fantoni et al. (1999, 2000) presented an energy based controller design strategy for stabilizing a pendubot. A hybrid design was developed for the pendubot by Zhang and Tarn (2002). Alvarez and Gallegos (2004) proposed a procedure to analyze the complex behavior of the pendubot and developed an unstable periodic orbits stabilizing controller. Li et al. (2004) presented an acrobatic control of pendubot with a fuzzy controller to keep the first link swinging periodically while the second link maintains standing vertically. Orlov et al. (2005) constructed a sliding-mode-like control design for underactuated systems with applications to the stabilization of a pendubot around the upright position to improve system robustness. Qian et al. (2007) suggested a hierarchical sliding mode control for the swing up of a pendubot. Albahkali et al. (2009) used the impulse momentum approach to swing up a pendubot. Shoji et al. (2010) simulated dexterous human throwing motion by the control of a pendubot supported by passive elements and some control via unstable zero dynamics.

In this chapter, the system dynamics of the pendubot is derived in the first section followed by the decoupling transformation and adaptive controller design. The last section presents the simulation results for an uncertain pendubot starting from various initial conditions.

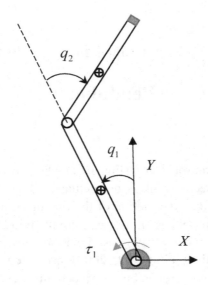

Figure 10.1. A pendubot

10.1 System Dynamics

Let q_1 and q_2 be the angular displacements of the lower link and the upper link. Then the kinetic energy of the pendubot can be found from the kinetic energy of the two links

$$K_{l_1} = \frac{1}{2}(I_1 + m_1 l_{c_1}^2)\dot{q}_1^2$$

$$K_{l_2} = \frac{1}{2}(I_2 + m_2 l_1^2 + m_2 l_{c_2}^2 + 2m_2 l_1 l_{c_2} \cos q_2)\dot{q}_1^2$$

$$+ (-I_2 - m_2 l_1 l_{c_2} \cos q_2 - m_2 l_{c_2}^2)\dot{q}_1\dot{q}_2 + \frac{1}{2}(I_2 + m_2 l_{c_2}^2)\dot{q}_2^2$$

where m_i, l_i and I_i are respectively the mass, length and moment of inertia of the *i*-th links, *i*=1,2. The total kinetic energy is the combination of K_{l_1} and K_{l_2} as

$$K = K_{l_1} + K_{l_2} = \frac{1}{2}(I_1 + m_1 l_{c_1}^2 + I_2 + m_2 l_1^2 + m_2 l_{c_2}^2 + 2m_2 l_1 l_{c_2} \cos q_2)\dot{q}_1^2$$
$$+ (-I_2 - m_2 l_1 l_{c_2} \cos q_2 - m_2 l_{c_2}^2)\dot{q}_1 \dot{q}_2 + \frac{1}{2}(I_2 + m_2 l_{c_2}^2)\dot{q}_2^2 \qquad (1)$$

With the definition of the potential energy

$$V = m_1 l_{c_1} g(\cos q_1 - 1) + m_2 l_1 g(\cos q_1 - 1) + m_2 l_{c_2} g[\cos(q_2 - q_1) - 1] \quad (2)$$

the Lagrangian can then be found as

$$L = \frac{1}{2}(\phi_1 + \phi_2 + 2\phi_3 \cos q_2)\dot{q}_1^2 - (\phi_2 + \phi_3 \cos q_2)\dot{q}_1 \dot{q}_2 + \frac{1}{2}\phi_2 \dot{q}_2^2$$
$$- \phi_4 g(\cos q_1 - 1) - \phi_5 g[\cos(q_2 - q_1) - 1] \qquad (3)$$

where we have defined $\phi_1 = I_1 + m_1 l_{c_1}^2 + m_2 l_1^2$, $\phi_2 = I_2 + m_2 l_{c_2}^2$, $\phi_3 = m_2 l_1 l_{c_2}$ and $\phi_4 = m_1 l_{c_1} + m_2 l_1$ and $\phi_5 = m_2 l_{c_2}$. The generalized coordinate vector and the generalized force vector can respectively be assigned as

$$\mathbf{q} = \begin{bmatrix} q_1 \\ q_2 \end{bmatrix} \in \Re^2$$

$$\boldsymbol{\tau} = \begin{bmatrix} u \\ 0 \end{bmatrix} \in \Re^2$$

Let us compute the quantities below to facilitate the derivation of the equation of motion via the Lagrange equation

$$\frac{\partial L}{\partial \dot{q}_1} = (\phi_1 + \phi_2 + 2\phi_3 \cos q_2)\dot{q}_1 - (\phi_2 + \phi_3 \cos q_2)\dot{q}_2$$

$$\frac{\partial L}{\partial q_1} = \phi_4 g \sin q_1 - \phi_5 g \sin(q_2 - q_1)$$

$$\frac{\partial L}{\partial \dot{q}_2} = \phi_2 \dot{q}_2 - (\phi_2 + \phi_3 \cos q_2)\dot{q}_1 \qquad (4)$$

$$\frac{\partial L}{\partial q_2} = -\phi_3 \sin q_2 \dot{q}_1^2 + \phi_3 \sin q_2 \dot{q}_1 \dot{q}_2 + \phi_5 g \sin(q_2 - q_1)$$

then we may have the dynamic equation of the cart pole system in the X-space using the Lagrange equation

$$\frac{d}{dt}\frac{\partial L}{\partial \dot{\mathbf{q}}} - \frac{\partial L}{\partial \mathbf{q}} = \tau \tag{5}$$

to have

$$(\phi_1 + \phi_2 + 2\phi_3 \cos q_2)\ddot{q}_1 - (\phi_2 + \phi_3 \cos q_2)\ddot{q}_2 - 2\phi_3 \sin q_2 \dot{q}_1 \dot{q}_2$$
$$+ \phi_3 \sin q_2 \dot{q}_2^2 - \phi_4 g \sin q_1 - \phi_5 g \sin(q_2 - q_1) = u \tag{6}$$
$$\phi_2 \ddot{q}_2 - (\phi_2 + \phi_3 \cos q_2)\ddot{q}_2^2 + \phi_3 \sin q_2 \dot{q}_1^2 - \phi_5 g \sin(q_2 - q_1) = 0$$

Define the state vector

$$\mathbf{x} = [q_1 \quad \dot{q}_1 \quad q_2 \quad \dot{q}_2]^T = [x_1 \quad x_2 \quad x_3 \quad x_4]^T \in \Re^4$$

and the state space representation becomes

$$\begin{aligned}
\dot{x}_1 &= x_2 \\
\dot{x}_2 &= f_2(\mathbf{x}) + b_2(\mathbf{x})u \\
\dot{x}_3 &= x_4 \\
\dot{x}_4 &= f_4(\mathbf{x}) + b_4(\mathbf{x})u
\end{aligned} \tag{7}$$

where

$$f_2(\mathbf{x}) = \frac{-\phi_2\phi_3 \sin x_3 (x_4 - x_2)^2 - \phi_3^2 \cos x_3 \sin x_3 x_2^2}{\phi_1\phi_2 - \phi_3^2 \cos^2 x_3}$$
$$+ \frac{\phi_2\phi_4 g \sin x_1 + \phi_3\phi_5 g \cos x_3 \sin(x_3 - x_1)}{\phi_1\phi_2 - \phi_3^2 \cos^2 x_3} \tag{8a}$$

$$b_2(x_3) = \frac{\phi_2}{\phi_1\phi_2 - \phi_3^2 \cos^2 x_3} \tag{8b}$$

$$f_4(\mathbf{x}) = \frac{(\phi_2 + \phi_3 \cos x_3)[\phi_4 g \sin x_1 - \phi_3 \sin x_3 (x_4 - x_2)^2]}{\phi_1\phi_2 - \phi_3^2 \cos^2 x_3}$$
$$+ \frac{(\phi_1 + \phi_3 \cos x_3)[\phi_5 g \sin(x_3 - x_1) - \phi_3 \sin x_3 x_2^2]}{\phi_1\phi_2 - \phi_3^2 \cos^2 x_3} \tag{8c}$$

$$b_4(x_3) = \frac{\phi_2 + \phi_3 \cos x_3}{\phi_1\phi_2 - \phi_3^2 \cos^2 x_3} \tag{8d}$$

There is only one control u appears in both the dynamic equations of the pendubot in (7) and hence the whole system is underactuated.

10.2 Coordinate Transformation

It is apparently that (7) is in the standard form (4.41) with $n = 4$. We may then transform the system from the X-space into Z-space by using the mapping (4.42) as

$$z_1 = x_1 - \int_0^{x_3} \frac{b_2(s)}{b_4(s)} ds = x_1 - \int_0^{x_3} \frac{\dfrac{\phi_2}{\phi_1\phi_2 - \phi_3^2 \cos^2 s}}{\dfrac{\phi_2 + \phi_3 \cos s}{\phi_1\phi_2 - \phi_3^2 \cos^2 s}} ds$$

$$= x_1 - \frac{2\phi_2}{\sqrt{\phi_2^2 - \phi_3^2}} \tan^{-1}\left[\sqrt{\frac{\phi_2 - \phi_3}{\phi_2 + \phi_3}} \tan\left(\frac{x_3}{2}\right)\right]$$

$$z_2 = x_2 - \frac{b_2(\mathbf{x})}{b_4(\mathbf{x})} x_4 = x_2 - \frac{\dfrac{\phi_2}{\phi_1\phi_2 - \phi_3^2 \cos^2 x_3}}{\dfrac{\phi_2 + \phi_3 \cos x_3}{\phi_1\phi_2 - \phi_3^2 \cos^2 x_3}} = x_2 - \frac{\phi_2}{\phi_2 + \phi_3 \cos x_3} x_4$$

$$z_3 = x_3$$

$$z_4 = x_4$$

With this transformation, the system dynamics in the Z-space becomes

$$\begin{aligned}
\dot{z}_1 &= z_2 + d_1(\mathbf{z}) \\
\dot{z}_2 &= z_3 + d_2(\mathbf{z}) \\
\dot{z}_3 &= z_4 + d_3(\mathbf{z}) \\
\dot{z}_4 &= d_4(\mathbf{z}) + b_4(\mathbf{z})u
\end{aligned} \tag{9}$$

where $d_1(\mathbf{z}) = d_3(\mathbf{z}) = 0$. Let $\theta = \sqrt{\dfrac{\phi_2 - \phi_3}{\phi_2 + \phi_3}} \tan\left(\dfrac{z_3}{2}\right)$, then the rest of the disturbances are defined as

$$d_2(\mathbf{z}) = -\frac{1}{\phi_2 + \phi_3 \cos z_3} \left\{ \phi_5 g \sin\left[z_3 - \left(z_1 + \frac{2\phi_2}{\sqrt{\phi_2^2 - \phi_3^2}} \tan^{-1}\theta \right) \right] \right.$$

$$\left. - \phi_3 \sin z_3 \left(z_2 + \frac{1}{\phi_2 + \phi_3 \cos z_3} \right)^2 \right\} - \frac{\phi_2 \phi_3 \sin z_3}{(\phi_2 + \phi_3 \cos z_3)^2} z_4$$

$$d_4(\mathbf{z}) = \frac{1}{\phi_1 \phi_2 - \phi_3^2 \cos^2 z_3} \left\{ (\phi_2 + \phi_3 \cos z_3) \left[\phi_4 g \sin\left(z_1 + \frac{2\phi_2}{\sqrt{\phi_2^2 - \phi_3^2}} \tan^{-1}\theta \right) \right.\right.$$

$$\left. - \phi_3 \sin z_3 \left[z_4 - \left(z_2 + \frac{1}{\phi_2 + \phi_3 \cos z_3} z_4 \right)^2 \right] \right]$$

$$+ (\phi_2 + \phi_3 \cos z_3) \left[\phi_5 g \sin\left(z_3 - z_1 - \frac{2\phi_2}{\sqrt{\phi_2^2 - \phi_3^2}} \tan^{-1}\theta \right) \right.$$

$$\left.\left. - \phi_3 \sin z_3 \left(z_2 + \frac{1}{\phi_2 + \phi_3 \cos z_3} z_4 \right)^2 \right] \right\}$$

$$b_4(z_3) = \frac{\phi_2 + \phi_3 \cos z_3}{\phi_1 \phi_2 - \phi_3^2 \cos^2 z_3}$$

10.3 Simulation Cases

The actual values of the system parameters in the simulation are selected as $m_1 = 0.55\,\text{kg}$, $m_2 = 0.23\,\text{kg}$, $l_1 = l_2 = 0.2\,\text{m}$, $l_{c1} = 0.1\,\text{m}$, $l_{c2} = 0.05\,\text{m}$, $I_1 = 0.00138\,\text{kg-m}^2$, and $I_2 = 0.00744\,\text{kg-m}^2$. The controller in (4-46) together with the desired trajectories (4-45) and the robust term (4-48) are used with the parameters selected as $c_i = 10, i = 1,...,4$. The first 21 terms of the Fourier series are used as the basis functions. The weighting matrices in the update law (4-50) are picked as $\Gamma_1 = \Gamma_2 = 0.1 I_{11}$ and $\Gamma_3 = \Gamma_4 = 0.01 I_{11}$. The constant σ in (4-50) is set to zero to turn off the σ-modification in the update law. The initial weighting in (4-50) are set to be $\hat{\mathbf{w}}_i(0) = [0 \quad 0 \quad \cdots \quad 0]^T \in \mathfrak{R}^{11}$, $i = 1,...,4$. Four simulation cases are presented here. The first case is the control of an uncertain inverted pendulum with $q_1(0) = 0.2$ and $q_2(0) = 0.5$. The rest of the cases are similar to the first case except different initial conditions to prove that the controller is able to stabilize the system with a large domain of attraction.

CASE 1: Uncertain pendubot with $q_1(0) = 0.2$ and $q_2(0) = 0.5$

Let us assume that initially the first link is at 0.2 rad. and the second link is at 0.5 rad. from rest and we would like to send them back to the vertical position, i.e., the origin. So, the initial state is $\mathbf{x} = [0.2 \quad 0 \quad 0.5 \quad 0]^T$ and the desired state is $\mathbf{x}_d = [0 \quad 0 \quad 0 \quad 0]^T$. The simulation results are shown in Figs. 10.2 to 10.4. Fig. 10.2 verifies that both links converge nicely regardless of the uncertainties in the system dynamics. Fig. 10.3 is the control effort required. Fig. 10.4 presents the convergence of each error surface.

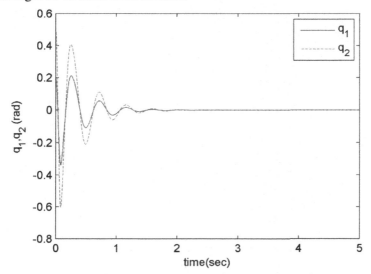

Figure 10.2. Trajectories of the links in case 1

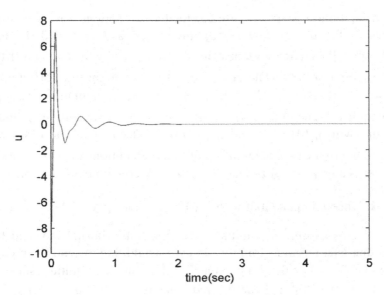

Figure 10.3.   Control effort

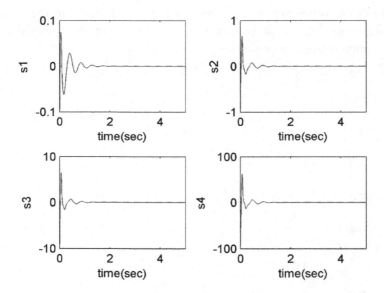

Figure 10.4.   Trajectories of error surfaces

CASE 2: Uncertain pendubot with $q_1(0) = 0.5$ and $q_2(0) = 0.5$

In this case, different initial positions of the two links are considered. It is much difficult to control due to the large initial angular displacement. The system model is still assumed to be unknown and the same set of controller parameters are used as in the previous case. The simulation results are shown in Figs. 10.5 to 10.7. Fig. 10.5 shows that the controller can still bring the system to the desired states in about 2 seconds which is almost the same as in the previous case regardless of the uncertainties in the system model. Fig. 10.6 is the history of the control effort which is still with a reasonable size. Fig. 10.7 is the convergence of each error surface.

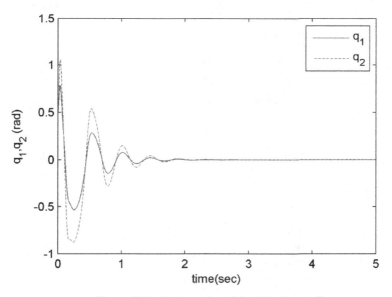

Figure 10.5. Trajectories of the links in case 2

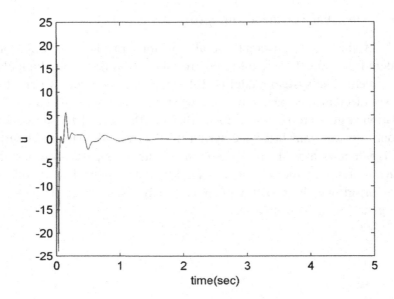

Figure 10.6.   Control effort

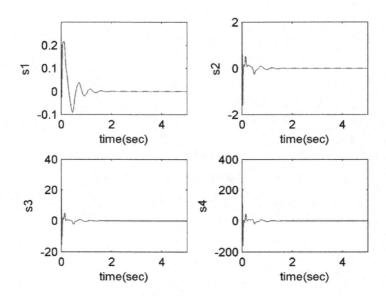

Figure 10.7.   Trajectories of error surfaces

CASE 3: Uncertain pendubot with $q_1(0) = -0.5$ and $q_2(0) = 0.5$

This set of initial conditions also makes the control difficult. The system model is still assumed to be not available and the same set of controller parameters as in the previous cases are used here. The simulation result is shown in Fig. 10.8 which indicates that the controller can still bring the system to the desired states in 2 seconds even with this set of initial conditions. Don't forget that the system model is not available during the control activity.

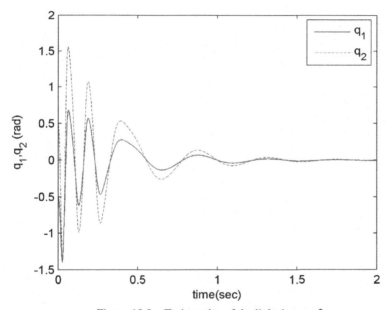

Figure 10.8.   Trajectories of the links in case 3

CASE 4: Uncertain pendubot with $q_1(0) = 1$ and $q_2(0) = 1$

This set of initial conditions makes the control more challenging because intuitively, in the present configuration, the control to benefit any one link might largely jeopardize the other. The same set of controller parameters are used here as in the previous cases. Fig. 10.9 shows that the controller can still bring the system to the desired states in

2 seconds even with this set of initial conditions regardless of the uncertainties in the system model.

Therefore, the same set of controller parameters is able to stabilize the pendubot in various initial conditions given in these four simulation cases with similar performance. Convergence of the two links for these cases can be observed to be rapid which confirms the transient performance derived in chapter 4. We may thus conclude that the coordinate transformation is useful and the control strategy is effective.

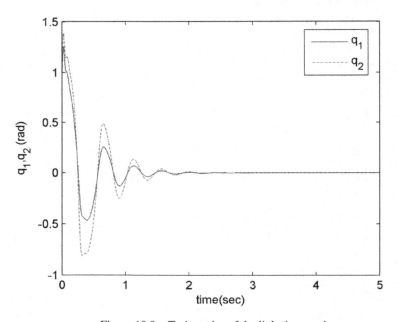

Figure 10.9. Trajectories of the links in case 4

Bibliography

1. E. D. Abdel-Rahman, A. H. Nayfeh, and Z. N. Masoud, "Dynamics and control of cranes: a review," *Journal of Vibration and Control*, vol. 9, no. 7, pp. 863-908, 2003.
2. M. Abé and T. Igusa, "Semi-active dynamic vibration absorbers for controlling transient response," *Journal of Sound and Vibration*, vol. 198, no. 5, pp. 547-569, 1996.
3. J. A. Acosta, "Furuta's pendulum: a conservative nonlinear model for theory and practice," *Mathematical Problems in Engineer*, Article ID 742894, 2010.
4. L. T. Aguilar, I. Boiko, L. Fridman and R. Iriarte, "Generating self-excited oscillations via two-relay controller," *IEEE Transactions on Automatic Control*, vol. 54, no. 2, pp. 416-420, Feb. 2009.
5. M. A. Ahmad, "Active sway suppression techniques of a gantry crane system," *European Journal of Scientific Research*, vol. 27, no. 3, pp. 322-333, 2009.
6. M. Alamir, "On friction compensation without friction model," *Proceedings IFAC 15th Triennial World Congress*, Barcelona, Spain, 2002.
7. T. Albahkali, R. Mukherjee and T. Das, "Swing-up control of the pendubot -- an impulse momentum approach," *IEEE Transactions on Robotics*, vol. 25, no. 4, pp. 975-982, August 2009.
8. A. Alleyne, "Physical insights on passivity-based TORA control designs," *IEEE Transactions on Control Systems Technology*, vol. 6, no. 3, pp. 436-439, May 1998.

9. H. Al-Shuka, B. Corves and W. H. Zhu, "Function approximation technique based adaptive virtual decomposition control for a serial chain manipulator," *Robotica*, pp. 1-25, Jan. 2014.

10. H. G. Alvarez and J. A. Gallegos "Experimental analysis and control of a chaotic pendubot," *The International Journal of Robotics Research*, vol. 23, no. 9, pp. 891-901, September 2004.

11. K. J. Astrom, J. Aracil and F. Gordillo, "A family of smooth controller for swinging up a pendulum," *Automatica*, vol. 36, no. 2, pp. 1841-1848, 2008.

12. N. Z. Azlan and H. Yamaura, "FAT based adaptive impedance control for unknown environment position," *World Academy of Science, Engineering and Technology*, vol. 6, pp. 1089-1094, Aug. 2012.

13. N. Z. Azlan and H. Yamaura, "Adaptive impedance control for unknown non-flat environment," *World Academy of Science, Engineering and Technology*, vol. 7, pp. 161-166, Aug. 2013.

14. A. El-Badawy, and T. N. El-Deen, "Quadratic nonlinear control of a self-excited oscillator," *Journal of Vibration and Control*, vol. 13, no. 4, pp. 403-414, 2007.

15. B. Balachandran, Y. Y. Li, and C. C. Fang, "A mechanical filter concept for control of nonlinear crane-load oscillations," *Journal of Sound and Vibration*, vol. 228, no. 3, pp. 651-682, 1999.

16. B. R. Barmish and G. Leitmann, "On ultimate boundedness control of uncertain systems in the absence of matching condition," *IEEE Transactions on Automatic Control*, vol. 27, no. 1, pp. 153-158, 1982.

17. G. Bartolini, A. Pisano and E. Usai, "Second-order sliding-mode control of container cranes," *Automatica*, vol. 38, no. 10, pp. 1783-1790, 2002.

18. M. J. Brennan and J. Dayou, "Global control of vibration using a tunable vibration neutralizer," *Journal of Sound and Vibration*, vol. 232, no. 3, pp. 585-600, 2000.

19. R. T. Bupp, D. S. Bernstein, V. S. Chellaboina and W. M. Haddad, "Resetting virtual absorbers for vibration control," *Journal of Vibration and Control*, vol. 6, no. 1, pp. 61-83, 2000.

20. R. T. Bupp, C. J. Wan, V. T. Coppola and D. S. Bernstein, "Design of a rotational actuator for global stabilization of translational motion," *Proceedings Symposium of Active Control of Vibration Noise*, ASME Winter Meeting, 1994.

21. R. A. Burdisso and J. D. Heilmann, "A new dual-reaction mass dynamic vibration absorber actuator for active vibration control," *Journal of Sound and Vibration*, vol. 214, no. 5, pp. 817-831, 1998.

22. T. Burg and D. Dawson, "Additional notes on the TORA example: a filtering approach to eliminate velocity measurements," *IEEE Transactions on Control Systems Technology*, vol. 5, no. 5, pp. 520-523, Sept. 1997.

23. A. I. Carlos, S. C. Miguel and O. Oscar, "The direct Lyapunov method for the stabilization of the Furuta pendulum," *International Journal of Control*, vol. 83, no. 11, pp. 2285-2293, Nov. 2010.

24. Y. C. Chang and J. Shaw, "Low-Frequency Vibration Control of a Pan/Tilt Platform with Vision Feedback," *Journal of Sound and Vibration*, vol. 302, Issue 4-5, pp. 716-727, May 2007.

25. Y. D. Chen, C. C. Fuh and P. C. Tung, "Application of voice coil motors in active dynamic vibration absorbers," *IEEE Transactions on Magnetics*, vol. 41, no. 3, pp. 1149-1154, 2005.

26. P. C. Chen and A. C. Huang, "Adaptive sliding control of active suspension systems based on function approximation technique," *Journal of Sound and Vibration*, vol. 282, issue 3-5, pp. 1119-1135, April 2005a.

27. P. C. Chen and A. C. Huang, "Adaptive multiple-surface sliding control of hydraulic active suspension systems based on function approximation technique," *Journal of Vibration and Control*, vol. 11, no. 5, pp. 685-706, 2005b.

28. P. C. Chen and A. C. Huang, "Adaptive sliding control of active suspension systems with uncertain hydraulic actuator dynamics," *Vehicle System Dynamics*, vol. 44, no. 5, pp. 357-368, May 2006.

29. Y. F. Chen and A. C. Huang, "Controller design for a class of underactuated mechanical systems," *IET Control Theory & Applications*, vol.6, Issue 1, pp. 103-110, 2012.

30. Y. F. Chen and A. C. Huang, "Adaptive control of rotary inverted pendulum system with time-varying uncertainties," *Nonlinear Dynamics*, vol. 76, pp. 95-102, 2014.

31. M. C. Chien and A. C. Huang, "Adaptive Impedance Control of Robot Manipulators based on Function Approximation Technique," *Robotica*, vol. 22, issue 04, pp. 395-403, August, 2004.

32. M. C. Chien and A. C. Huang, "Adaptive control of flexible-joint electrically-driven robot with time-varying uncertainties," *IEEE Transactions on Industrial Electronics*, vol. 54, no. 2, pp. 1032-1038, April 2007.

33. M. C. Chien and A. C. Huang, "An adaptive controller design for flexible-joint electrically-driven robots with consideration of time-varying uncertainties," Chapter 5 in the book *Frontiers in Adaptive Control*, I-Tech Education and Publishing, Vienna, Austria, 2009.

34. M. C. Chien and A. C. Huang, "A regressor-free adaptive control for flexible-joint robots based on function approximation technique," Chapter 2 in the book *Advances in Robot Manipulators*, I-Tech Education and Publishing, Vienna, Austria, 2010a.

35. M. C. Chien and A. C. Huang, "Design of a FAT-based adaptive visual servoing for robots with time-varying uncertainties," *International Journal of Optomechatronics*, vol. 4, Issue 2, pp. 93-114, 2010b.

36. M. C. Chien and A. C. Huang, "Adaptive impedance controller design for flexible-joint electrically-driven robots without computation of the regressor matrix," *Robotica*, vol. 30, pp. 133-144, 2012.

37. K. Cohen, R. Yaffe, T. Weller and J. Z. Ben-Asher, "Experimental studies on adaptive fuzzy control of a smart structure," *Journal of Vibration and Control*, vol. 8, no. 8, pp. 1071-1083, 2002.

38. J. Collado, R. Lozano and I. Fantoni, "Control of convey-crane based on passivity," *Proceedings of American Control Conference*, pp. 1260-1264, 2000.

39. S. Cong, Y. Liang and W. Shang, "Function approximation based sliding mode adaptive control for time-varying uncertain nonlinear systems," in the book *Frontiers in Adaptive Control*, I-Tech Education and Publishing, Vienna, Austria, 2009.

40. G. Corriga, A. Giua and G. Usai, "An implicit gain-scheduling controller for cranes," *IEEE Transactions on Control Systems Technology*, vol. 6, no. 1, pp. 15-20, 1998.

41. C. L. Davis and G. A. Lesieutre, "An actively tuned solid-state vibration absorber using capacitive shunting of piezoelectric stiffness," *Journal of Sound and Vibration*, vol. 232, no. 3, pp. 601-617, 2000.

42. Y. Fang, W. E. Dixon, D. M. Dawson and E. Zergeroglu, "Nonlinear coupling control laws for an underactuated overhead crane system," *IEEE/ASME Transactions on Mechatronics*, vol. 8, no. 3, pp. 418-423, 2003.

43. I. Fantoni, R. Lozano, M. W. Spong, "Passivity based control of the pendubot," *Proceedings American Control Conference*, San Diego, California, June 1999.

44. I. Fantoni, R. Lozano, and M. W. Spong, "Energy based control of the pendubot," *IEEE Transactions on Automatic Control*, vol. 45, no. 4, pp. 725-729, April 2000.

45. I. Fantoni and R. Lozano, *Nonlinear Control for Underactuated Mechanical Systems*, London: Springer Verlag, 2002.

46. I. Fantoni and R. Lozano, "Stabilization of the Furuta pendulum around its homoclinic orbit," *International Journal of Control*, vol. 75, no. 6, pp. 390-398, 2002.

47. D. Filipović and D. Schröder, "Band-pass vibration absorber," *Journal of Sound and Vibration*, vol. 214, no. 3, pp. 553-566, 1998.

48. D. Filipovic and D. Schroder, "Vibration absorption with linear active resonators: continuous and discrete time design and analysis," *Journal of Vibration and Control*, vol. 5, no. 5, pp. 685-708, 1999.

49. L. Freidovich, A. Shiriaev, F. Gordillo, F. Gomez-Estern and J. Aracil, "Partial energy shaping control for orbital stabilization of high frequency oscillations of the Furuta pendulum," *IEEE Transactions on Control Systems Technology*, vol. 17, no. 4, pp. 853-858, July 2009.

50. K. Furuta, M. Yamakita and S. Kobayashi, "Swing-up control of inverted pendulum," *Processings International Conference on Industrial Electronics*, Control and Instrumentation, pp. 2193-2198, 1991.

51. S. S. Ge, C. C. Hang, T. H. Lee and T. Zang, *Stable Adaptive Neural Network Control*, Boston: Kluwer Academic, 2001.

52. J. C. Gerdes and J. K. Hedrick, "Vehicle speed and spacing control via coordinated throttle and brake actuation," *Control Engineering Practice*, vol. 5, no. 11, pp. 1607-1614, 1997.

53. J. C. Gerdes and J. K. Hedrick, "Loop-at-a-time design of dynamic surface controller for nonlinear systems," *Proceedings of American Control Conference*, pp. 3574-3578, 1999.

54. F. Gordillo, J. A. Acosta and J. Aracil, "A new swing-up law for the Furuta pendulum," *International Journal of Control*, vol. 76, no. 8, pp. 836-844, 2003.

55. J. H. Green and J. K. Hedrick, "Nonlinear speed control for automotive engines," *Proceedings of American Control Conference*, pp. 2891-2897, 1990.

56. J. P. den Hartog, *Mechanical Vibrations*, New York: McGraw-Hill, 1956.

57. L. Hera, L. B. Freidovich, A. S. Shiriaev and U. Mettin, "New approach for swing up the Furuta pendulum: theory and experiments," *Mechatronics*, vol. 19, pp. 1240-1250, July 2009.

58. J. Hollkamp and T. Starchville, "A self-tuning piezoelectric vibration absorber," *Journal of Intelligent Material Systems and Structures*, vol. 5, pp. 559-566, 1994.

59. M. Hosek, N. Olgac and H. Elmali, "The centrifugal delayed resonator as a tunable torsional vibration absorber for multi-degree-of-freedom systems," *Journal of Vibration and Control*, vol. 5, no. 2, pp. 299-322, 1999.

60. A. C. Huang and Y. C. Chen, "Adaptive sliding control for single-link flexible-joint robot with mismatched uncertainties," *IEEE Transactions on Control Systems Technology*, vol.12, no.5, pp. 770-775, 2004a.

61. A. C. Huang and Y. C. Chen, "Adaptive multiple-surface sliding control for non-autonomous systems with mismatched uncertainties," *Automatica*, vol. 40, issue 11, pp. 1939-1945, Nov. 2004b.

62. A. C. Huang and Y. S. Kuo, "Sliding control of nonlinear systems containing time-varying uncertainties with unknown bounds," *International Journal of Control*, vol. 74, no. 3, pp. 252-264, 2001.

63. A. C. Huang and K. K. Liao "FAT-based adaptive sliding control for flexible arms, theory and experiments," *Journal of Sound and Vibration*, vol. 298, issue 1-2, pp. 194-205, Nov. 2006.

64. A. C. Huang, S. C. Wu and W. F. Ting, "An FAT-based adaptive controller for robot manipulators without regressor matrix: theory and experiments," *Robotica*, vol. 24, pp. 205-210, 2006.

65. A. C. Huang and M. C. Chien, *Adaptive Control of Robot Manipulators – A Unified Regressor Approach*, World Scientific, 2010.

66. S. Huang and R. Lian, "A dynamic absorber with active vibration control," *Journal of Sound and Vibration*, vol. 178, no. 3, pp. 323-335, 1994.

67. L. C. Hung, H. P. Lin, and H. Y. Chung, "Design of self-tuning fuzzy sliding mode control for TORA system," *Expert Systems with Applications*, vol. 32, pp. 201-212, 2007.

68. S. Huyanan and N. D. Sims, "Vibration control strategies for proof mass actuators," *Journal of Vibration and Control*, vol. 13, no. 12, pp. 1785-1806, 2007.

69. P. A. Ioannou and J. Sun, *Robust Adaptive Control*, Prentice Hall, 1996.

70. A. Isidori, *Nonlinear Control Systems*, 2nd ed., Springer-Verlag, 1989.

71. M. Jankovic, D. Fontaine and P. V. Kokotovic, "TORA example: cascade- and passivity-based control design," *IEEE Transactions on control systems technology*, vol. 4, no. 3, pp. 292-297, May 1996.

72. N. Jalili and D. W. Knowles, "Structural vibration control using an active resonator absorber: modeling and control implementation," *Smart Materials and Structures*, vol. 13, pp. 998-1005, 2004.

73. C. Y. Kai and A. C. Huang, "A regressor-free adaptive controller for robot manipulators without Slotine and Li's modification," *Robotica*, vol. 31, Issue 7, pp. 1051-1058, Oct. 2013a.

74. C. Y. Kai and A. C. Huang, "Linearization of rate-dependent nonlinearity with a compensator in feedback configuration," *Mechanical Systems and Signal Processing*, vol. 39, pp. 333-341, 2013b.

75. C. Y. Kai and A. C. Huang, "Adaptive control of brushless DC motors without model reduction," *Applied Mechanics and Materials*, Special Issue on Advances in Mechatronics and Control Engineering, vol. 278-280, pp. 1409-1412, 2013c.

76. C. Y. Kai and A. C. Huang, "A regressor-free adaptive impedance controller for robot manipulators without Slotine and Li's modification: theory and experiments," accepted by *Robotica*, 2014a.

77. C. Y. Kai and A. C. Huang, "Active vibration absorber design for mechanical systems with frequency-varying excitations," accepted by *Journal of Vibration and Control*, 2014b.

78. H. K. Khalil, *Nonlinear Systems*, 2nd ed., Prentice Hall, 1996.

79. P. V. Kokotovic, "The joy of feedback: nonlinear and adaptive," IEEE Control Systems Magazine, pp. 7- 17, June 1992.

80. P. V. Kokotovic and R. A. Freeman, "Design of softer robust nonlinear control laws," *Automatica*, vol. 29, no. 6, pp. 1425-1437, 1993.

81. J. H. Koo, M. Ahmadian, M. Setareh and T. Murray, "In search of suitable control methods for semi-active tuned vibration absorbers," *Journal of Vibration and Control*, vol. 10, no. 2, pp. 163-174, 2004.

82. M. Krstic, I. Kanellakopoulos, and P. Kokotovic, Nonlinear and *Adaptive Control Design*, John Wiley and Sons, 1995.

83. H. H. Lee, "A new design approach for the anti-swing trajectory control of overhead cranes with high-speed hoisting," *International Journal of Control*, vol. 77, no. 10, pp. 931-940, 2004.

84. C. H. Lee, H. H. Chang and B. H. Wang, "Decoupled adaptive type-2 fuzzy controller design for nonlinear TORA systems," *Proceeding of IEEE Fuzzy Conference*, pp. 506-512, July 16-21, 2006.

85. C. H. Lee, S. K. Chang, "Experimental implementation of nonlinear TORA system and adaptive backstepping controller design," *Neural Computation Application*, vol. 21, pp. 785-800, 2012.

86. T. F. Lee and A. C. Huang, "Vibration suppression in belt-driven servo systems containing uncertain nonlinear dynamics," *Journal of Sound and Vibration*, vol. 330, Issue 1, pp. 17-26, 2011.

87. G. Leitmann and Y. H. Chen, "Robustness of uncertain systems in the absence of matching assumptions," *International Journal of Control*, no. 45, pp. 1527-1542, 1987.

88. J. Li, H. X. Hua and R. Y. Shen, "Saturation-based active absorber for a non-linear plant to a principal external excitation," *Mechanical Systems and Signal Processing*, vol. 21, pp. 1489-1498, 2007.

89. J. Li, B. Shen, F. Bai, C. M. Chew and C. L. Teo, "First implementation results on FAT based adaptive control for a lower extremity rehabilitation device," *Proceedings IEEE International Conference on Mechatronics and Automation*, Takamatsu, Japan, August 4-7, 2013.

90. W. Li, K. Tanaka, and H. O. Wang, "Acrobatic control of a pendubot," *IEEE Transactions on Fuzzy Systems*, vol. 12, no. 4, August 2004.

91. C. Li, J. Yi and D. Zhao, "Control of the TORA system using SIRMs based type-2 fuzzy logic," *Proceedings IEEE International Conference on Fuzzy Systems*, Jeju Island, Korea, pp. 694–699, 2009.

92. Y. Liang, S. Cong and W. Shang, "Function approximation based sliding mode adaptive control," *Nonlinear Dynamics*, vol. 54, pp. 223-230, 2008.

93. Y. Liu and H. Yu, "A survey of underactuated mechanical systems," *IET Control Theory and Applications*, vol. 7, Issue 7, pp. 921-935, 2013.

94. Z. N. Masoud, M. F. Daqaq and N. A. Nayfeh, "Pendulation reduction on small ship mounted telescopic cranes," *Journal of Sound and Vibration*, vol. 10, no. 8, pp. 1167-1179, 2004.

95. D. H. McMahon, J. K. Hedrick and S. E. Shladover, "Vehicle modeling and control for automated highway systems," *Proceedings American Control Conference*, pp. 297-303, 1990.

96. R. A. Morgan and K. W. Wang, "Active-passive piezoelectric absorbers for systems under multiple non-stationary harmonic excitations," *Journal of Sound and Vibration*, vol. 255, no. 4, pp. 685-700, 2002.

97. K. Nagaya, A. Kurusu, S. Ikai and Y. Shitani, "Vibration control of a structure by using a tunable absorber and an optimal vibration absorber under auto-tuning control," *Journal of Sound and Vibration*, vol. 228, no. 4, pp. 773-792, 1999.

98. S. Nair, "A normal form for energy shaping: application to the Furuta pendulum," *Proceedings IEEE Conference on Decision and Control*, vol. 1, pp. 516-521, Dec. 2002.

99. K. S. Narendra and A. M. Annaswamy, *Stable Adaptive Systems*, Prentice Hall, 1989.

100. S. Nazrulla and H. K. Khalil, "A novel nonlinear output feedback control applied to the TORA benchmark system," *Proceedings IEEE Conference on Decision and Control*, Cancun, 2008.

101. R. W. O'Flaherty, R. G. Sanfelice and A. R. Teel, "Hybrid control strategy for robust global swing-up of the pendubot," *Proceedings American Control Conference*, Seattle, Washington, USA June 11-13, 2008.

102. K. Ogata, *Modern Control Engineering*, 3rd edition, Prentice Hall, 1997.

103. D. C. D. Oguamanam, J. S. Hansen and G. R. Heppler, "Dynamics of a three-dimensional overhead crane system," *Journal of Sound and Vibration*, vol. 242, no. 3, pp. 411-426, 2001.

104. R. Olfati-Saber and A. Megretasi, "Controller design for a class of underactuated nonlinear systems," *Proceedings of IEEE Conference on Decision and Control*, pp. 4182-4187, Tampa FL, Dec. 1998.

105. R. Olfati-Saber, "Cascade normal form for underactuated mechanical systems," *Proceedings IEEE Conference on Decision and Control*, vol. 3, pp. 2162-2167, Dec. 2000.

106. R. Olfati-Saber, *Nonlinear Control of Underactuated Mechanical Systems with Application to Robotics and Aerospace Vehicles*, Dissertation, Department of Electrical Engineering and Computer Science, MIT, Feb. 2001.

107. R. Olfati-Saber, "Normal forms for underactuated mechanical systems with symmetry," *IEEE Transactions on Automatic Control*, vol. 47, no. 2, pp. 305-308, 2002.

108. N. Olgac and B. Holm-Hansen, "A novel active vibration absorption technique: Delayed resonator," *Journal of Sound and Vibration*, vol. 176, pp. 93-104, 1994.

109. N. Olgac and N. Jalili, "Modal analysis of flexible beams with delayed resonator vibration absorber: theory and experiments," *Journal of Sound and Vibration*, vol. 218, no. 2, pp. 307-331, 1998.

110. N. Olgac and H. Elmali, "Analysis and design of delayed resonator in discrete domain," *Journal of Vibration and Control*, vol. 6, no. 2, pp. 273-289, 2000.

111. Y. Orlov, L. Aguilar and L. Acho, "Zeno mode control of underactuated mechanical systems with application to pendubot stabilization around the upright position," *16th IFAC World Congress*, Prague, Czech Republic, 2005.

112. J. Ormondroyd and J. P. Den Hartog, "The theory of the dynamic vibration absorber," *ASME Journal of Applied Mechanics*, vol. 50, pp. 9-22, 1928.

113. S. S. Oueini, A. H. Nayfeh and J. R. Pratt, "A nonlinear vibration absorber for flexible structures," *Nonlinear Dynamics*, vol. 15, pp. 259-282, 1998.

114. S. S. Oueini, and A. H. Nayfeh, "Analysis and application of a nonlinear vibration absorber," *Journal of Vibration and Control*, vol. 6, no. 7, pp. 999-1016, 2000.

115. M. S. Park and D. Chwa, "Swing-up and stabilization control of inverted pendulum systems via coupled sliding mode control method," *IEEE Transactions on Industrial Electronics*, vol. 56, no. 9, pp. 3541-3555, Sept. 2009.

116. Z. Petres, P. L. V´arkonyi, P. Baranyi, and P. Korondi,"Different Affine Decomposition of the Model of the TORA System by TP model transformation," *Proceedings International Conference on Intelligent Engineering Systems*, pp. 105-110, Mediterranean Sea, September 16–19, 2005.

117. D. Qian, J. Yi, and D. Zhao, "Hierarchical sliding mode control to swing up a pendubot," *Proceedings American Control Conference*, New York, pp. 5254–5259, 2007.

118. R. Rana and T. T. Soong, "Parametric study and simplified design of tuned mass dampers," *Engineering Structures*, vol. 20, no. 3, pp. 193-204, 1998.

119. C. E. Rohrs, L. S. Valavani, M. Athans and G. Stein, "Robustness of continuous time adaptive control algorithm in the presence of un-modeled dynamics," *IEEE Transactions on Automatic Control*, vol. 30, pp. 881-889, 1985.

120. S. E. Semercigil, D. Lammers and Z. Ying, "A new tuned vibration absorber for wide-band excitations," *Journal of Sound and Vibration*, vol. 156, no. 3, pp. 445-459, 1992.

121. A. S. Shiriaev, L. B. Freidovich, A. Robertsson, R. Johansson and A. Sandberg, "Virtual holonomic constraints based design of stable oscillations of Furuta pendulum: theory and experiments," *IEEE Transactions on Robotics and Automation*, vol. 23, no. 4, pp. 827-832, Aug. 2007.

122. T. Shoji, S. Nakaura, and M. Sampei, "Throwing motion control of the springed Pendubot via unstable zero dynamics," *Proceedings IEEE International Conference on Control Applications*, Yokohama, Japan, pp. 1602–1607, 2010.

123. K. K. Shyu, C. L. Jen and L. J. Shang, "Sliding-mode control for an underactuated overhead crane system," *Proceedings IEEE Conference on Industrial Electronics*, pp. 412-417, 2006.

124. J. J. E. Slotine and W. Li, *Applied Nonlinear Control*, Prentice Hall, Englewood Cliffs, New Jersey, 1991.

125. M. W. Spong, "The swing up control problem for the acrobot," *IEEE Control Systems Magazine*, vol. 15, pp. 49-55, 1995.

126. M. W. Spong and D. J. Block, "The pendubot: a mechanical system for control research and education," *Proceedings IEEE Conference on Decision and Control*, pp. 555-556, New Orleans, Dec. 1995.

127. M. W. Spong, "Energy based control of a class of underactuated mechanical systems," *Proceedings IFAC World Congress*, July 1996.

128. M. W. Spong and L. Praly, "Control of underactuated mechanical systems using switching and saturation," *Proceedings Block Island Workshop on Control Using Logic Based Switching*, 1996.

129. M. W. Spong, "Underactuated mechanical systems," In B. Sciliano and K. P. Valavanis (eds): *Control Problem in Robotics and Automation,* Springer Verlag, London, UK. 1997.

130. M. W. Spong, "Applications of switching control in robot locomotion," *Proceedings Workshop on Intelligent Control in robotics and Automation, IFAC World Congress*, Beijing, China, July 1999a.

131. M. W. Spong, "Passivity-based control of the compass gait biped," *Proceedings Workshop on Intelligent Control in robotics and Automation, IFAC World Congress*, Beijing, China, July 1999b.

132. M. W. Spong, P. Corke and R. Lozano, "Nonlinear control of the inertia wheel pendulum," *Automatica*, vol. 37, pp. 1845-1851, 1999.

133. J. T. Spooner, M. Maggiore, R. Ordonez and K. M. Passino, *Stable Adaptive Control and Estimation for Nonlinear Systems – Neural and Fuzzy Approximator Techniques*, NY: John Wiley & Sons, 2002.

134. I. Stakgold and M. Holst, *Green's Functions and Boundary Value Problems*, 3rd ed., Wiley, 2011.

135. A. Stotsky, J. K. Hedrick and P. P. Yip, "The use of sliding modes to simplify the backstepping method," *Proceedings of American Control Conference*, pp. 1703-1708, 1997.

136. V. Sukontanakarn and M. Parnichkun, "Real-time optimal control for rotary inverted pendulum," *American Journal of Applied Sciences*, vol. 6, pp. 1106-1115, 2009.

137. J. Q. Sun, M. R. Jolly and M. A. Norris, "Passive, adaptive and active tuned vibration absorbers — a survey," *ASME Journal of Mechanical Design*, vol. 117, pp. 234-242, 1995.

138. H. L. Sun, P. Q. Zhang, X. L. Gong and H. B. Chen, "A novel kind of active resonator absorber and the simulation on its control effort," *Journal of Sound and Vibration*, vol. 300, pp. 117-125, 2007.

139. B. Talaei, F. Abdollahi, H. A. Talebi, and E. O. Karkani, "Whole arm manipulation planning based on feedback velocity fields and sampling based techniques," *ISA Transactions*, vol. 53, pp. 684-691, 2013.

140. G. Tadmor, "Dissipative design and the nonlinear TORA benchmark example, revisited," *Proceedings IEEE Conference on Decision and Control*, Phoenix, Arizona USA, December, 1999.

141. G. Tadmor, "Dissipative design and lossless dynamics, and the nonlinear TORA benchmark example," *IEEE Transactions on Control System Technology*, vol. 9, no. 2, pp. 391-398, March 2001.

142. K. Tanaka, T. Taniguchi and H. O. Wang, "Fuzzy control based on quadratic performance function - a LMI approach," *Proceedings IEEE Conference on Decision and Control*, Tampa, Florida USA, December, 1998.

143. Y. C. Tsai and A. C. Huang, "FAT based adaptive control for pneumatic servo system with mismatched uncertainties," *Mechanical Systems and Signal Processing*, vol.22, no.6, pp. 1263-1273, 2008a.

144. Y. C. Tsai and A. C. Huang, "Multiple-surface sliding controller design for pneumatic servo systems," *Mechatronics*, no. 18, pp. 506-512, Nov. 2008b.

145. T. Turker, H. Gorgun and G. Cansever, "Lyapunov's direct method for stabilization of the Furuta pendulum," *Turkish Journal of Electrical Engineering & Computer Sciences*, vol. 120, no. 1, pp. 99-110, 2012.

146. F. Tyan and S. C. Lee, "An adaptive control of rotating stall and surge of jet engine - a function approximation approach," *IEEE Conference on Decision and Control and 2005 European Control Conference*, pp. 5498-5503, Dec. 2005.

147. N. Uchiyama, "Robust control of rotary crane by partial-state feedback with integrator," *Mechatronics*, vol. 19, no. 8, pp. 1294-1302, 2009.

148. P. L. Walsh and J. S. Lamancusa, "A variable stiffness vibration absorber for minimization of transient vibrations," *Journal of Sound and Vibration*, vol. 158, no. 2, pp. 195-211, 1992.

149. C. J. Wan, D. S. Bernstein and V. T. Coppola, "Global stabilization of the oscillating eccentric rotor," *Proceedings IEEE Conference on Decision and Control*, pp. 4024-4029, Dec. 1994.

150. K. Williams, G. Chiu and R. Bernhard, "Adaptive passive absorbers using shape memory alloys," *Journal of Sound and Vibration*, vol. 249, no. 5, pp. 835-848, 2002.

151. M. Won and J. K. Hedrick, "Multiple-surface sliding control of a class of uncertain nonlinear systems," *International Journal of Control*, vol. 64, no. 4, pp. 693-706, 1996.

152. R. I. Wright and M. R. F. Kidner, "Vibration absorbers: a review of applications in interior noise control of propeller aircraft," *Journal of Vibration and Control*, vol. 10, no. 8, pp. 1221-1237, 2004.

153. J. H. Yang and K. S. Yang, "Adaptive coupling control for overhead crane systems," *Mechatronics*, vol. 17, no. 2-3, pp. 143-152, 2007.

154. P. P. Yip and J. K. Hedrick, "The use of sliding modes to simplify the backstepping control method," *Proceedings American Control Conference*, pp. 1703-1708, 1997.

155. C. H. Yu, F. C. Wang, and Y. J. Lu, "Robust control of a Furuta pendulum," *SICE Annual Conference*, pp. 2559-2563, Aug. 2010.

156. M. Zhang and T. J. Tarn, "Hybrid control of the pendubot," *IEEE/ASME Transactions on Mechatronics*, vol. 7, no. 1, pp. 79-86, March 2002.

Index

Printed in the United States
By Bookmasters